T0205660

MECHANICAL AND PHYSICO-CHEMICAL CHARACTERISTICS OF MODIFIED MATERIALS

Performance Evaluation and Selection

MECHANICAL AND PHYSICO-CHEMICAL CHARACTERISTICS OF MODIFIED MATERIALS

Performance Evaluation and Selection

Edited by
Seghir Maamir, PhD
A. K. Haghi, PhD

Apple Academic Press Inc. | Apple Academic Press Inc.
3333 Mistwell Crescent | 9 Spinnaker Way
Oakville, ON L6L 0A2 | Waretown, NJ 08758
Canada | USA

© 2016 by Apple Academic Press, Inc.

First issued in paperback 2021

Exclusive worldwide distribution by CRC Press, a member of Taylor & Francis Group

No claim to original U.S. Government works

ISBN 13: 978-1-77463-239-0 (pbk)
ISBN 13: 978-1-77188-092-3 (hbk)

Library of Congress Control Number: 2015946544

Library and Archives Canada Cataloguing in Publication

Mechanical and physico-chemical characteristics of modified materials: performance evaluation and selection / edited by Seghir Maamir, PhD, A.K. Haghi, PhD.

Includes bibliographical references and index.
Issued in print and electronic formats.
ISBN 978-1-77188-092-3 (hardcover).--ISBN 978-1-4987-1410-5 (ebook)
1. Materials--Mechanical properties. 2. Materials--Testing.
3. Chemistry, Physical and theoretical. I. Haghi, A. K., author, editor II. Maamir, Seghir, editor

---

TA410.M43 2015   620.1'10287   C2015-905183-5   C2015-905222-X

Apple Academic Press also publishes its books in a variety of electronic formats. Some content that appears in print may not be available in electronic format. For information about Apple Academic Press products, visit our website at **www.appleacademicpress.com** and the CRC Press website at **www.crcpress.com**

CONTENTS

    Poly(3-hydroxybutyrate) – Polyvinyl Alcohol Blends 191

    A. A. Ol'khov, A. L. Iordanskii, and G. E. Zaikov

9. Prospective Production Ways of New Heat-Resisting
    Materials Based on Polyimides 205

    E. T. Krut'ko and N. R. Prokopchuk

10. Protective Properties of Modified Epoxy Coatings 215

    N. R. Prokopchuk, M. V. Zhuravleva,N. P. Ivanova, T. A. Zharskaya,
    and E. T. Krut'ko

11. Modification of Polyamide-6 by N,N'-bis-Maleamidoacid 225

    N. R. Prokopchuk, E. T. Krut'ko, and M. V. Zhuravleva

12. Efficient Synthesis of a
    Bisspiro-3,1-Benzoxazin-4,1'-Cyclopentanes 237

    Sh. M. Salikhov, R. R. Zaripov, S. A. Krasko, L. R. Latypova,
    N. M. Gubaidullin, and I. B. Abdrakhmanov

13. Polyconjugated Linear and Cardo Oligomers Based on
    Aromatic Diamines and Their Thermal Properties 249

    A. F. Yarullin, L. E. Kuznetsova, A. F. Yarullina, Kh. S. Abzaldinov,
    O. V. Stoyanov, and G. E. Zaikov

14. Alkylation of Aromatic Hydrocarbons by
    Polyhalogen-Containing Cyclopropanes 257

    A. N. Kazakova, G. Z. Raskildina, N. N. Mikhaylova, L. V. Spirikhin,
    and S. S. Zlotsky

15. GnRH Analogs As Modulator of LH and FSH:
    Exploring Clinical Importance 269

    Debarshi Kar Mahapatra, Vivek Asati, and Sanjay Kumar Bharti

    Index 317

LIST OF CONTRIBUTORS

I. B. Abdrakhmanov
Institute of Organic Chemistry Ufa Scientific Centre of Russian Academy of Sciences, Prospect Oktyabrya 71, 450054, Ufa, Russia

M. I. Abdullin
Bashkir State University, Ufa, 450077, Russia, E-mail: ProfAMI@yandex.ru

Kh. S. Abzaldinov
Kazan National Research Technological University, Kazan, Russia

D. S. Andreev
Volgograd State Architect-build University, Sebrykov Department, Russia

Vivek Asati
School of Pharmaceutical Sciences, Guru Ghasidas Vishwavidyalaya (A Central University), Bilaspur–495009, Chhattisgarh, India

V. A. Babkin
Volgograd State Architect-build University, Sebrykov Department, Russia

A. A. Basyrov
Bashkir State University, Ufa, 450077, Russia, E-mail: ProfAMI@yandex.ru

Sanjay Kumar Bharti
School of Pharmaceutical Sciences, Guru Ghasidas Vishwavidyalaya (A Central University), Bilaspur–495009, Chhattisgarh, India

Ali Fazlipur
Department of Mechanical Engineering, Imam Hossein Comprehensive University, Tehran, Iran; E-mail: fazlipourali1368@yahoo.com

A. B. Glazyrin
Bashkir State University, Ufa, 450077, Russia, E-mail: ProfAMI@yandex.ru

N. M. Gubaidullin
Bashkir State Agrarian University, 50 Let Oktyabrya, 21, 450001, Ufa, Russia, Tel: +7 (347) 235 55 60

A. K. Haghi
Department of Textile Engineering, University of Guilan, Rasht, Iran; E-mail: AKHaghi@yahoo.com

A. L. Iordanskii
Semenov Institute of Chemical Physics, Russian Academy of Sciences, Kosygin str. 4, Moscow, 119991 Russia

N. P. Ivanova
Assistant Professor (BSTU), Belarusian State Technological University, Sverdlova Str.13a, Minsk, Republic of Belarus

S. G. Karpova
Emanuel Institute of Biochemical Physics, Russian Academy of Sciences, 4 Kosygina str., 119991 Moscow, Russia

A. N. Kazakova
Ufa State Petroleum Technological University, Kosmonavtov Str. 1, 450062 Ufa, Russia; Tel: +(347) 2420854, E-mail: nocturne@mail.ru

Hossein Khodarahmi
Department of Mechanical Engineering, Imam Hossein Comprehensive University, Tehran, Iran

S. A. Krasko
Ufa State Petroleum Technological University, Kosmonavtov 1, 450062, Ufa, Russia, Tel: + 7 (347) 242 09 35. E-mail: ksa.85@mail.ru

E. T. Krut'ko
Professor (BSTU), Belarusian State Technological University, Sverdlova Str.13a, Minsk, Republic of Belarus

O. S. Kukovinets
Bashkir State University, Ufa, 450077, Russia, E-mail: ProfAMI@yandex.ru

L. E. Kuznetsova
Kazan National Research Technological University, Kazan, Russia

L. R. Latypova
Institute of Organic Chemistry Ufa Scientific Centre of Russian Academy of Sciences, Prospect Oktyabrya 71, 450054, Ufa, Russia

N. M. Livanova
Emanuel Institute of Biochemical Physics, Russian Academy of Sciences, 4 Kosygina str., 119991 Moscow, Russia; E-mail: livanova@sky.chph.ras.ru

Debarshi Kar Mahapatra
School of Pharmaceutical Sciences, Guru Ghasidas Vishwavidyalaya (A Central University), Bilaspur–495009, India; Tel.: +91 7552–260027; Fax: +91 7752–260154; E-mail: mahapatradebarshi@gmail.com

N. N. Mikhaylova
Ufa State Petroleum Technological University, Kosmonavtov Str. 1, 450062 Ufa, Russia

B. Hadavi Moghadam
Department of Textile Engineering, University of Guilan, Rasht, Iran

D. A. Nguyen
Kazan National Research Technological University, 420015, Kazan, K. Marx str., 68, Russia

A. A. Ol'khov
Plekhanov Russian University of Economics, Stremyanny per. 36, Moscow117997 Russia, E-mail: aolkhov72@yandex.ru

A. A. Popov
Emanuel Institute of Biochemical Physics, Russian Academy of Sciences, 4 Kosygina str., 119991 Moscow, Russia

K. Yu. Prochukhan
Bashkir State University, Kommunisticheskaya ul., 19, Ufa, Respublika Bashkortostan, 450076, Russia

A. Yu. Prochukhan
Bashkir State University, Kommunisticheskaya ul., 19, Ufa, Respublika Bashkortostan, 450076, Russia

N. R. Prokopchuk
Professor (BSTU), Corresponding Member of Belarus NAS, Belarusian State Technological University, Sverdlova Str.13a, Minsk, Republic of Belarus; E-mail: v.polonik@belstu.by

G. Z. Raskildina
Ufa State Petroleum Technological University, Kosmonavtov Str. 1, 450062 Ufa, Russia

Sh. M. Salikhov
Institute of Organic Chemistry Ufa Scientific Centre of Russian Academy of Sciences, Prospect Oktyabrya 71, 450054, Ufa, Russia, Tel: +7 (347) 235 55 60. E-mail: Salikhov@anrb.ru

L. V. Spirikhin
Institute of Organic Chemistry, Ufa Scientific Center, Russian Academy of Sciences, Oktyabrya Avenue 71, 450054 Ufa, Russia

I. A. Starostina
Kazan National Research Technological University, 420015, Kazan, K. Marx str., 68, Russia

O. V. Stoyanov
Kazan National Research Technological University, 420015, Kazan, K. Marx str., 68, Russia, E-mail: ov_stoyanov@mail.ru

D. V. Vezenov
Lehigh University, 27 Memorial Dr. W. Bethlehem, PA 18015, USA

A. F. Yarullina
Kazan National Research Technical University Named After A.N. Tupolev, Kazan, Russia; E-mail: aleksej-yarullin@yandex.ru, abzaldinov@mail.ru

G. E. Zaikov
N.M. Emanuel Institute of Biochemical Physics, Russian Academy of Sciences, 4 Kosygin str., Moscow 119334; Kazan National Research Technological University, Kazan, Russia, E-mail: chembio@sky.chph.ras.ru

R. R. Zaripov
Bashkir State Agrarian University, 50 Let Oktyabrya, 21, 450001, Ufa, Russia, Tel: +7 (347) 235 55 60

T. A. Zharskaya
Assistant Professor (BSTU), Belarusian State Technological University, Sverdlova Str.13a, Minsk, Republic of Belarus

M. V. Zhuravleva
PhD student (BSTU), Belarusian State Technological University, Sverdlova Str.13a, Minsk, Republic of Belarus

S. S. Zlotsky
Ufa State Petroleum Technological University, Kosmonavtov Str. 1, 450062 Ufa, Russia

LIST OF ABBREVIATIONS

2D	two-dimensional
3D	three-dimensional
BBF	best bin first
CAD	computer-aided design
CCD	charged coupled device
CM	Chamfer matching
CMT	correct-match rate
DCM	directional chamfer matching
DLT	direct linear transform
DOF	degrees of freedom
DoG	difference of Gaussians
DSM	digital surface models
DTCWT	dual-tree complex wavelet transforms
EBR	edge-based regions
EBSD	electron backscatter diffraction
ENMs	electrospun nanofibrous membranes
ESM	efficient second-order minimization method
FN	false negatives
FP	false positives
GIS	geographic information system
GLOH	gradient location and orientation histogram
GSD	ground sampling distance
HD	Hausdorff distance
HOG	histogram of oriented gradient
IBR	intensity-extrema-based
LMS	least median of squares
LoG	Laplacian of Gaussian
LSCM	laser scanning confocal microscope
Micro-CT	micro computed tomography
MSER	maximally stable extremal regions

NN	nearest neighbors
OD	optic disk
PCA	principal component analysis
POA	pore open area
QMF	quadrature mirror filter
SEM	scanning electron microscopy
SIFT	scale invariant feature transform
SSD	sum of squared differences
SURF	speeded up robust features
SVD	singular value decomposition of a matrix
TN	true negatives
TP	true positives
TPR	true positive rate
VP	vanishing point
WDST	windowed discriminant spectral template

LIST OF SYMBOLS

ε	porosity
V_s	volume of sample
V_p	pore volume
K_{KC}	Kozeny-Carman predicted permeability, mD
c	a constant
d	median grain size diameter, microns
P_n	projective space (n-dimensions)
A_n	affine Space (n-dimensions)
R^{n+1}	vector space
$X = [X_1, X_1, ..., X_{n+1}]^T$	homogeneous coordinates
L	line in projective space
π	plane in projective space
S^2	2D sphere
$f(x) = Ax + b$	affine transformations
A	square matrix
b	translation matrix
$G(x, \sigma\%)$	Gaussian matrix
C	Harris detector matrix
λ	Eigen values
$\sigma\%$	natural scale
H	Hessian matrix
I	image
I_{xx}, I_{yy}, I_{xy}	second order derivatives of image intensity
$f(x, y)$	two-dimensional image function
$*$	discrete convolution
$g(x, y)$	filter kernel
$x-y$	direction of a Guassian
$M(x, y)$	image gradient magnitude
$Q(x, y)$	image orientation
$h_{r(l,m)}(k)$	Gradient magnitude

c_k	orientation bin center
Δ_k	orientation bin width
$H(X,\sigma)$	Hessian matrix
$L_{xx}(X,\sigma)$	convolution of the Gaussian second order derivative
$H(C)$	finite energy
ρ_b	DTCWT coefficients
α and β	scaling coefficients
g_k	individual feature in gist descriptor
$w_k(x, y)$	A spatial window
m_{ij}	distance ground between pairs of features across the two images
c_{ij}	cost of matching these two points
$h_i(k), h_j(k)$	K-bin normalized histogram at pi and qj
$H(A, B)$	Hausdorff distance
$S = \{S_1 = \pm 1, ..., s_N\}$	binary sequences
$d_{CM}(U,V)$	chamfer distance between U and V
$W(x,s)$	a warping function
t_x, t_y	translations along x and y axis
$\hat{s}$	alignment parameter
$\varphi(x)$	direction term
λ	a weighting factor between location and orientation terms
$H(M,R)$	Hausdorff distance between M and R (M and R are reference feature points and image feature points)
$\| \ \|$	distance between two points
$K^{th}_{a[A}$	K^{th} ranked value of $d_B(a)$
$d_B(a)$	minimum distance value at point a to the point set B
$Q^{th}_{b[B}$	K^{th} ranked value of the Euclidean distance set
$P^{th}_{a[A}$	P^{th} ranked value of $Q^{th}_{b[B}\|a{-}b\|$
NE	size of the Euclidean distance
p*	true nearest neighbor
$M = [X, Y, Z]^T$	a 3D point
$m\%$	homogeneous coordinate vector of vector m

List of Symbols

K	camera calibration matrix
e_1, e_2	epipoles
l_1, l_2	epipolar lines
U, V	orthogonal 3×3 matrices
Σ	3×3 diagonal matrix
R	rotation matrix
H	homogeneous matrix
E	essential matrix
F	fundamental matrix
N	number of iterations
[	probability that a sample correspondence
Q_i	rotation vector
$P_i\% P_j\%$	camera rays
t_{ij}	translation vector between camera centers

PREFACE

Understanding chemical and solid materials and their properties and behavior is fundamental to chemical design and engineering design and is a key application of chemicals and materials science. Written for all students of chemical science and mechanical engineering and materials science and design, this book describes the procedures for material selection and design in order to ensure that the most suitable materials for a given application are identified from the full range of materials, chemicals, and section shapes available.

Several case studies have been developed to further illustrate procedures and to add to the practical implementation of the text.

This new volume reviews recent academic and technological developments behind new engineered modified materials. The book is intended for researchers and those interested in future developments in mechanical and physico-chemical characteristics of modified materials. Several innovative applications for different materials are described in considerable detail with emphasis on the experimental data that supports these new applications. From fibers to chemical materials and from membranes to ceramics, creative modifications concerning new composites are described that could one day become commonplace. Never before has this much new information materials modification been packaged into one volume. In this book the world's leading experts describe their most recent research in their areas of expertise. The book will also be a useful tool for students and researchers, providing helpful insights into new evolving research areas in mechanical and physico-chemical characteristics of modified materials.

ABOUT THE EDITORS

Seghir Maamir, PhD
Seghir Maamir, PhD, is an Associate Professor at the Department of Mechanical Engineering of the Faculty of Engineer Sciences at the University of Boumerdes, Algeria. He holds a PhD from the University of Franche Comte (France). He was examiner of magister thesis in polymers and composites of the University of Boumerdes, Algeria. His fields of interest are fluid mechanics, heat transfer and mechanical vibrations.

A. K. Haghi, PhD
A. K. Haghi, PhD, is the Member of the Canadian Research and Development Center of Sciences and Cultures (CRDCSC), Montreal, Quebec, Canada; and Editor-in-Chief, *International Journal of Chemoinformatics and Chemical Engineering*; and Editor-in-Chief, *Polymers Research Journal*. He holds a BSc in urban and environmental engineering from the University of North Carolina (USA); a MSc in mechanical engineering from North Carolina A&T State University (USA); a DEA in applied mechanics, acoustics and materials from the Université de Technologie de Compiègne (France); and a PhD in Engineering Sciences from the Université de Franche-Comté (France). He is the author and editor of 165 books as well as 1000 published papers in various journals and conference proceedings. Dr. Haghi has received several grants, consulted for a number of major corporations, and is a frequent speaker to national and international audiences. Since 1983, he served as a professor at several universities.

CHAPTER 1

SIMULATION OF PENETRATION OF STEEL PROJECTILE WITH OGIVE NOSE INTO THE SANDWICH TARGET WITH FOAM, POLYSTYRENE AND POLY RUBBER CORE USING AUTODYNE SOFTWARE

ALI FAZLIPUR and HOSSEIN KHODARAHMI

Department of Mechanical Engineering, Imam Hossein Comprehensive University, Tehran, Iran; E-mail: fazlipourali1368@ yahoo.com

CONTENTS

ABSTRACT

In this chapter, the penetration of steel projectile with Ogive nose into the sandwich target with foam, polystyrene and poly rubber core is simulated using Autodyne software. The behavioral equation used in the simulation is Johnson Cook for projectile and metal, Johnson Holmquist for ceramic, and crushable foam for foam. Also the equation of state used in steel and metal projectile is linear, in ceramic layer is polynomial, and in polystyrene and poly rubber is shock. The simulation was performed using Lagrangian approach. In each model, the output speed of projectile and ballistic limit velocity is calculated and the impact of different foam, polystyrene and poly rubber cores on ballistic resistance of sandwich panel is investigated.

1.1   INTRODUCTION

Polymer matrix composites are the most common composites include matrices made of polymer (resin), which are connected to phase distribution amplifier. Polymeric materials, such as, epoxy and polyester do not have very high mechanical properties compared to metals. Reinforcing fibers in polymer matrix composites should have mechanical properties, including very high tensile strength, than the polymer matrix. By increasing reinforcing polymer matrix, the tensile strength and elastic modulus of the composite will increase. While applying load the bulk of force will be tolerated by fibers and indeed polymer matrix protects fibers from physical and chemical damages, in addition to transmitting the force to fibers. In addition the matrix holds fibers together like a glue and of course it limits the crack spread [1]. Cho et al. [2] experimentally investigated experimental tests by drop apparatus on closed-cell aluminum foam and compared them with the results of the simulation. Griskevicius et al. [3] experimentally studied the behavior of honeycomb cored sandwich structures under quasi-static and dynamic loading. Ruan et al. [4] experimentally investigated mechanical behavior and energy absorption of aluminum foam sandwich panels under quasi-static step loads. In this study, the penetration of steel projectile with Ogive, flat and hemispherical nose, with a speed of 854 m/s into the sandwich target is simulated and the effect of projectile nose shape on the output speed of the target is studies.

The analysis method in this study is based on the Lagrangian viewpoint. This view is based on following the path of material particle movement. In fact, this method is used to obtain the history of material movement during analysis. Applying this method, the solving environment (of the material) is divided into a series of elements. Nodal points of the elements are connected to the material and move by changing the material shape [5].

1.2   SIMULATION OF PENETRATION INTO SANDWICH PANEL

Autodyne software (version 12) is the product of Dynamic Century Co. and its presentation to engineering community dates back to 1986. Multipurpose software applications are designed so that they can analyze high-rate engineering issues. These analyzes are performed by using finite difference and finite volume methods in a wide range of nonlinear problems in solids, liquids and gases dynamics. Issues that usually have such specifications are highly function of time and have nonlinear geometrical factors such as very large deformations [5].

1.2.1   CHARACTERISTICS OF TARGET AND PROJECTILE

As shown in Fig. 1.1, the steel projectile of 28.7 mm length and 7.62 mm diameter collides to the sandwich panel of 600 mm diameter at a speed of 854 to 554 m/s. The simulated cores in Fig. 1.2 consist of a foam, polystyrene and polyethylene rubber.

FIGURE 1.1   Projectile with Ogive nose.

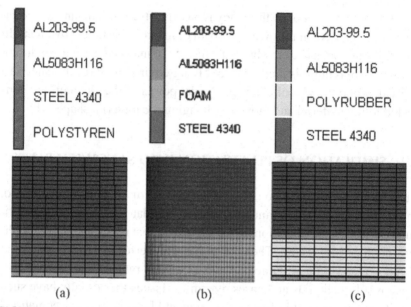

FIGURE 1.2   Element classification of the target: (a) poly rubber core, (b) foam core, (c) polystyrene core.

Projectile and target modeling has done in an axial symmetrical way. In part, the sandwich panel and projectile were simulated by selecting Lagrangian method. Surrounding environment of the target was fully bounded. Lagrangian element classification of projectile and target and target arrangement is shown in Fig. 1.2.

1.2.2 MATERIAL MODELS

Material models of Johnson-Cook are commonly used for metals under the impact loading and high-rate strains. In this model, the flow stress in independent terms are considered as a function of plastic strain, strain rate and temperature. One of the requirements of using this model is using the equation of state. Johnson and Cook have been suggested following equation to show the material flow stress.

$$\sigma_y = (A + B\bar{\varepsilon}^{p^n})(1 + C\ln\dot{\varepsilon}^*)(1 - T^{*m}) \qquad (1)$$

where σ_y is the flow stress, $\bar{\varepsilon}^{p''}$ is effective plastic strain, $\dot{\varepsilon}^*$ is effective plastic strain rate, T^* is dimensionless temperature and A, B, C, n and m are material constants. In this material model, failure strain is obtained according to the following equation in terms of dimensionless stress σ^*, strain rate ε^* and temperature T^* and material constants D1 to D5 values:

$$\sigma^* = \frac{p}{\sigma_{eff}} \qquad (2)$$

Failure occurs when the damage coefficient is equals to 1 [9]:

$$D = \sum \frac{\Delta\bar{\varepsilon}^p}{\varepsilon^f} \qquad (3)$$

Coefficients related to aluminum sheet are shown in Table 1.1.

1.2.3 JOHNSON–HOLMQUIST MATERIAL MODEL

Johnson–Holmquist material model is used for concrete, ceramic, glass and other brittle materials. The equivalent stress in these materials is the function of degradation factor D as follows:

$$\sigma^* = \sigma^i - D\left(\sigma_i^* - \sigma_f^*\right) \qquad (4)$$

where

$$\sigma_i^* = a\left(p^* + t^*\right)^n\left(1 + c\,ln\,\dot{\varepsilon}^*\right) \qquad (5)$$

TABLE 1.1 Material Coefficients of Johnson–Cook for Aluminum Layer of the Target [8]

Equation of State	Linear	Strength	Johnson-Cook
Reference density	0.27×10^1 (KPa)	Shear Modulus	2.69×10^7 (KPa)
Bulk Modulus	5.83×10^7 (J/mKs)	Yield Stress	1.67×10^5 (KPa)
Reference Temperature	300 (K)	Hardening Constant	5.96×10^5 (KPa)
Specific Heat	910 (J/mKs)	Erosion	Geometric Strain
Thermal Conductivity	Geometric Strain	Erosion Strain	1 (none)

Eq. (4) shows the behavior of not destroyed materials. The sign * indicates normalized quantities. Normalized tensile stress is obtained by dividing equivalent tensile stress to elastic limit stress of Hogoniot (HEL) according to Eq. (5). Normalized pressure is also obtained by dividing equivalent pressure to Hogoniot elastic limit (HEL) according to Eqs. (2)–(7). In above equations, c, is the strength parameter for different strain rates and $\dot{\varepsilon}^*$ is the normalized plastic strain rate.

$$t^* = \frac{T}{P_{HEL}} \tag{6}$$

$$P^* = \frac{T}{P_{HEL}} \tag{7}$$

Degradation coefficient is also as follows.

$$D = \sum \frac{\Delta \varepsilon^p}{\varepsilon_f^p} \tag{8}$$

where $\Delta \varepsilon_f^p$ is plastic strain development and ε_f^p is fracture strain, which is defined as follows

$$\varepsilon_f^p = d_1 \left(P^* + t^* \right)^{d_2} \tag{9}$$

where d_1 and d_2 are coefficients, which are determined by the user. The following equation shows the behavior of damaged materials:

$$\sigma_f^* = b \left(P^* \right)^m \left(1 + c \, ln \, \dot{\varepsilon}^* \right) \leq sfma \tag{10}$$

In the above equation, b is the coefficient imposed by the user and *sfma* is the normalized maximum fracture strength. d_1 is the controller of amount rate and if it equals to zero, complete degradation occurs in a time interval. In non-damaged materials hydrostatic pressure is as follows:

$$P = k_1 \mu + k_2 \mu^2 + k_3 \mu^3 \tag{11}$$

where

$$\mu = \frac{\rho}{\rho_0} - 1 \tag{12}$$

In this equation ρ_0 is the material density and ρ is the material density after deformation. Also according to Hogoniot elastic limit and shear modulus (G), the modulus μ_{HEL} is obtained using the following equation:

$$HEL = k_1\mu_{HEL} + k_2\mu_{HEL}^2 + k_3\mu_{HEL}^3 + (4/3)g(\mu_{HEL}/(1+\mu_{HEL})) \qquad (13)$$

By normalizing the above equation, Eq. (14) is obtained [9]:

$$P_{HEL} = k_1\mu_{HEL} + k_2\mu_{HEL}^2 + k_3\mu_{HEL}^3 \qquad (14)$$

And also

$$\sigma_{HEL} = 1.5\left(HEL - P_{HEL}\right) \qquad (15)$$

Coefficients of steel projectile, aluminum, foam, polystyrene and poly rubber are shown in Tables 1.2 and 1.3.

TABLE 1.2 Material Coefficients of Ceramic [7]

Equation of State	Polynomial	Strength	Johnson-Holmquist
Reference density	3.89 (g/cm3)	Shear Modulus	1.52×10^8 (KPa)
Bulk Modulus A1	2.31×10^8 (KPa)	Strain Rate Constant C	0.007 (none)
Parameter A2	-1.6×10^8 (KPa)	Intact Strength Constant A	0.88 (none)
Parameter A3	2.774×10^9 (KPa)	Erosion	Geometric Strain
Parameter B0	0 (KPa)	Erosion Strain	0.5(none)

TABLE 1.3 Material Coefficients of Foam [8]

Equation of State	Linear	Strength	Crushable Foam
Reference density	0.486 (g/cm^3)	Shear Modulus	1.923×10^4 (KPa)
Bulk modulus	4.166×10^4 (KPa)	Max Tensile Stress	4.79×10^3 (KPa)
Reference Temperature	293 (K)	Erosion	Geometric Strain
Specific Heat	0 (J/mKs)	Erosion Strain	1 (none)

1.3   RESULTS AND DISCUSSION

1.3.1   FOAM CORED SANDWICH PANEL

As shown in Table 1.4, the projectile penetrated into the sandwich target at the speed of 854 m/s and passed through it at 500 m/s. Then the projectile speed was reduced to 100 m/s, so that it exited the target at 368.4 m/s. In order to obtain the ballistic limit velocity, the deceleration trend was continued until the projectile collided to the model at 603 m/s and exited the target at 124 m/s. By decreasing the output speed, the initial velocity would decrease and the projectile hit the target at 602 m/s and stopped in it. The velocity of 602.5 m/s was calculated as the ballistic limit velocity.

TABLE 1.4   Results of Numerical Simulation

Target	Initial speed (m/s)	Output speed (m/s)	Ballistic limit speed (m/s)
Foam cored sandwich panel	854	500	
	754	368.4	
	654	193.5	
	627	172.8	602.5
	605	132	
	603	124	
	602	0	
Polystyrene cored sandwich panel	854	476.8	
	754	354.2	
	654	231	622.5
	625	32	
	620	0	
Poly rubber cored sandwich panel	854	498.3	
	754	366.3	
	654	277.6	612.5
	615	17.49	
	610	0	

1.3.2 POLYSTYRENE CORED SANDWICH PANEL

As shown in Table 1.4, the projectile penetrated into the sandwich target at the speed of 854 m/s and passed through it at 476.8 m/s. Then the projectile speed was reduced to 100 m/s, so that it exited the target at 354.2 m/s. In order to obtain the ballistic limit velocity, the deceleration trend was continued until the projectile collided to the model at 625 m/s and exited the target at 32 m/s. By decreasing the output speed, the initial velocity would decrease and the projectile hit the target at 620 m/s and stopped in it. The velocity of 622.5 m/s was calculated as the ballistic limit velocity.

1.3.3 POLYURETHANE RUBBER CORED SANDWICH PANEL

As shown in Table 1.4, the projectile penetrated into the sandwich target at the speed of 854 m/s and passed through it at 498.3 m/s. Then the projectile speed was reduced to 100 m/s, so that it exited the target at 366.3 m/s. In order to obtain the ballistic limit velocity, the deceleration trend was continued until the projectile collided to the model at 615 m/s and exited

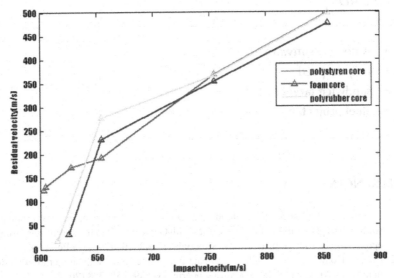

FIGURE 1.3   Diagram of initial velocity according to output speed.

the target at 17.46 m/s. By decreasing the output speed, the initial velocity would decrease and the projectile hit the target at 610 m/s and stopped in it. The velocity of 612.5 m/s was calculated as the ballistic limit velocity (Fig. 1.3).

1.4 CONCLUSION

In this research, the penetration of steel projectile with Ogive nose into the sandwich target with foam, polystyrene and poly rubber core is simulated using Autodyne software. Behavioral equation used in the simulation is Johnson Cook for projectile and metal, Johnson Holmquist for ceramics, and Crushable foam for foam, which show the behavior of material well by having related coefficients. The simulation is performed by using Lagrangian approach. In each model, the output speed of projectile and ballistic limit velocity is calculated and it is shown that due to the penetration of projectile with Ogive nose into the foam cored sandwich target, ballistic limit velocity has the lowest value, and in sandwich panel with polystyrene core it has the highest value.

KEYWORDS

- Autodyne software
- output speed
- sandwich target
- steel projectile

REFERENCES

1. Haghighat, M., Vahedi Kh., Mehdi Pur Omrani, A. Analytical and experimental study of full penetration of the projectiles with Ogive nose into the composite carbon/epoxy targets. Journal of Energetic Materials, 17, 43–50 (1391) (in Persian).
2. Cho, J., Hong, S., Lee, S., Cho, C. Impact fracture behavior at the material of aluminum foam, Materials Science and Engineering A, 539, 250–258 (2012).

3. Griskevicius, P., Zeleniakiene, D., Lfisis, V., Ostrowski, M. Experimental and Numerical Study of Impact Energy Absorption of Safety Important Honeycomb Core Sandwich Structures. Materials Science, 16, 119–123 (2010).
4. Ruan, D., Sridhar, I., Lu G. R., Wong, Y. C. Quasi-static indentation tests on aluminum foam sandwich panels, Composite Structures, 94, 1745–1754 (2012).
5. Saedi Daryan, A., Jalili, S. Blast and Impact Engineering with AUTODYNE applications. Tehran publisher, Darian engineers Publications, (1391) (in Persian).
6. Shafiei, M., Vahedi Kh. Analytical and numerical study of long rod projectile penetration into concrete and concrete – steel targets., Journal of Energetic Materials, 8, 43–51 (1392) (in Persian).
7. Chi R., Serjouei, A., Sridhar, I., Tan, G. Ballistic impact on bi-layer alumina/aluminum armor: A semianalytical approach. International Journal of Impact Engineering, 52, 37–46 (2013).
8. Hou, W., Zhu, F., Lu, G., Fang, D-N. Ballistic impact experiments of metallic sandwich panels with aluminum foam core. International Journal of Impact Engineering, 37, 1045–1055 (2010).

CHAPTER 2

SIMULATION OF PENETRATION OF TUNGSTEN PROJECTILE INTO THE COMPOSITE CERAMICS – METAL TARGET USING AUTODYNE SOFTWARE

ALI FAZLIPUR and HOSSEIN KHODARAHMI

Department of Mechanical Engineering, Imam Hossein Comprehensive University, Tehran, Iran; E-mail: fazlipourali1368@ yahoo.com

CONTENTS

ABSTRACT

In this chapter, the penetration of tungsten projectile into the composite ceramics – metal target is stimulated using Autodyne software. The behavioral equation used in this simulation is Johnson Cook for projectile and metal and Johnson Holmquist for ceramic. Also, the equation of state used in tungsten projectile and metal is linear and in ceramic layer is Polynomial. In each model the output speed of the projectile is calculated and compared with the results of experimental tests. The results obtained indicate a good agreement between performed simulation and experimental results. Then Kevlar-epoxy and glass-epoxy polymer layer add to the ceramic/ aluminum target and projectile penetration resistance is investigated.

2.1   INTRODUCTION

Ceramics, due to having properties such as low density, high compressive strength and high bulk and shear modulus, has long been used as a protective armor in ballistic systems. In applications, ceramic armors are usually reinforced by the supporter. The supporter absorbs the remaining kinetic energy after ceramic failure and stabilizes it in the penetration process. The more the thickness of backing sheet be, the less the ceramic failure would be due to the deformation of backing sheet. After the projectile hitting a ceramic target with metal backing layer, cone failure would occur due to the return of tensile waves. The ceramic cone passes the load from the projectile hitting to a wider surface of the cone base on the backing layer. Investigating the phenomena of penetration into ceramics is usually done in three ways: (i) Experimental and quasi-experimental methods, (ii) numerical methods, (iii) analytical or engineering methods. So far, various models ranging from numerical, experimental and analytical models in penetration of ceramic targets is presented. In numerical method, all equations governing penetration is completely solved by using different methods, such as finite difference, SPH and finite element in continuous environment. Numerical methods are appropriate for the analysis of complex and complicated issues [1]. Rosenberg et al. [2] examined and simulated long-rod penetrators in thick ceramic sandwich targets, between two steel sheets in laboratory. They showed that Johnson Holmquist failure model is a good model having few independent parameters, which can be used in empirical data. Sánchez Galvez et al. [3] considered a summary of numerical and analytical calculations of the ceramic/ metal and ceramic/composite during armor failure in order to get an optimal design. They concluded that adding ceramics and composite can be applied to the conventional armor protection of steel and aluminum. Prakash et al. [4] numerically investigated the effects of binder thickness at high speeds and performance of ceramic/aluminum targets. Madhu et al. [5] examined ballistic performance of ceramic targets of 95% and 99.5% with aluminum backing layer, with speeds of 500 to 830 m/s by vertically steel projectiles. Ning et al. [6] considered dynamic responses of ceramic targets by high-speed long rod tungsten projectiles to develop an analytical model. Tan et al. [7] considered the failure mechanisms of ceramic AD95/ steel armor 4340 under projectile penetration, with vertical speed of 820 m/s.

In this research tungsten projectile penetration into ceramic/metal target was first studied and the results of the simulation were compared with experimental test results and a relatively good agreement between results of simulation and experimental test was shown. Then in order to increase ballistic resistance of panels, Kevlar-epoxy and glass-epoxy polymer layers were added to the model and the projectile output speed were studied in each model. Among them, adding a layer of Kevlar-epoxy increased ballistic resistance of the structure to the penetration and decreased the speed of the projectile.

2.2 SIMULATION OF PENETRATION INTO CERAMIC – ALUMINUM TARGET

Autodyne software (version12) is the product of Dynamic century Co. and its presentation to engineering community dates back to 1986. Multipurpose software applications are designed so that they can analyze high-rate engineering issues. These analyzes are performed by using finite difference and finite volume methods in a wide range of nonlinear problems in solids, liquids and gases dynamics. Issues that usually have such specifications are highly function of time and have nonlinear geometrical factors such as very large deformations. Because most high- rate mechanical problems are associated with very large deformations, one of the crucial factors in preparation of issues for analysis is to observe this factor. In order to analyze engineering problems, various methods have been proposed so far. In this section numerical methods are based on finite element, finite difference, boundary element, and not networked methods [8].

2.2.1 CHARACTERISTICS OF TARGET AND PROJECTILE

The characteristics of modeled ceramics – aluminum target is shown in Table 2.1.

Projectile and target modeling has done in an axial symmetrical way. In part (part), ceramic and aluminum were simulated with Lagrangian methods and projectile was simulated in the form of SPH elements. The size of elements used in ceramics – aluminum target is 1 mm. Surrounding environment of the target was fully bound. Element classification of the target is shown in Fig. 2.1.

2.2.2 MATERIAL MODELS

Material models of Johnson-Cook are commonly used for metals under the impact loading and high-rate strains. In this model, the flow stress in independent terms are considered as a function of plastic strain, strain rate

TABLE 2.1   Characteristics of Projectile and Target

Materials	Density (kg/m3)	Diameter (mm)	Length (mm)	Speed (m/s)
Ceramic	3890	200	-	-
Aluminum	2700	200	-	-
Projectile	17,600	12	20	1240

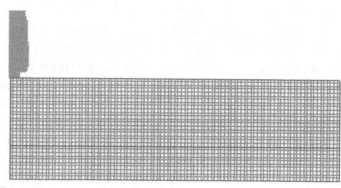

FIGURE 2.1   Geometry of projectile and target.

and temperature. One of the requirements of using this model is using the equation of state. Johnson and Cook have been suggested following equation to show the material flow stress.

$$\sigma_y = (A + B\bar{\varepsilon}^{p^n})(1 + C \ln \dot{\varepsilon}^*)(1 - T^{*m}) \tag{1}$$

where σ_y is the flow stress, $\bar{\varepsilon}^{p^n}$ is effective plastic strain, $\dot{\varepsilon}^*$ is effective plastic strain rate, T^* is dimensionless temperature and A, B, C, n and m are material constants. In this material model, failure strain is obtained according to the following equation in terms of dimensionless stress σ^*, strain rate ε^* and temperature T^* and material constants D1 to D5 values:

$$\sigma^* = \frac{p}{\sigma_{eff}} \tag{2}$$

Failure occurs when the damage coefficient is equals to 1 [9]:

$$D = \sum \frac{\Delta\bar{\varepsilon}^p}{\varepsilon^f} \tag{3}$$

Coefficients related to tungsten projectile and aluminum sheet are shown in Tables 2.2 and 2.3.

TABLE 2.2 Material Coefficients Tungsten Projectile [10]

Equation of State	Linear	Strength	Johnson-Cook
Bulk Modulus	3.1×10^8 (KPa)	Shear Modulus	2.69×10^7 (KPa)
Specific Heat	134 (J/mKs)	Yield Stress	1.67×10^5 (KPa)
Thermal Conductivity	0 (J/mKs)	Hardening Constant	5.96×10^5 (KPa)
Erosion	Geometric Strain	Erosion Strain	1 (none)

TABLE 2.3 Material Coefficients Aluminum Sheet [10]

Equation of State	Linear	Strength	Johnson-Cook
Bulk Modulus	3.1×10^7 (KPa)	Shear Modulus	1.6×10^7 (KPa)
Specific Heat	310 (J/mKs)	Yield Stress	1.67×10^5 (KPa)
Thermal Conductivity	910 (J/mKs)	Hardening Constant	5.96×10^5 (KPa)
Erosion	0 (J/mKs)	Erosion Strain	1 (none)

2.2.3 JOHNSON–HOLMQUIST MATERIAL MODEL

Johnson–Holmquist material model is used for concrete, ceramic, glass and other brittle materials. The equivalent stress in these materials is the function of degradation factor D as follows:

$$\sigma^* = \sigma^i - D\left(\sigma_i^* - \sigma_f^*\right) \tag{4}$$

where

$$\sigma_i^* = a\left(p^* + t^*\right)^n \left(1 + c\,ln\,\dot{\varepsilon}^*\right) \tag{5}$$

Eq. (4) shows the behavior of not destroyed materials. The sign * indicates normalized quantities. Normalized tensile stress is obtained by dividing equivalent tensile stress to elastic limit stress of Hogoniot (HEL) according to Eq. (5). Normalized pressure is also obtained by dividing equivalent pressure to hogoniot elastic limit (HEL) according to Eqs. (2)–(7). In above equations, c, is the strength parameter for different strain rates and $\dot{\varepsilon}^*$ is the normalized plastic strain rate.

$$t^* = \frac{T}{P_{HEL}} \tag{6}$$

$$P^* = \frac{T}{P_{HEL}} \tag{7}$$

Degradation coefficient is also as follows.

$$D = \sum \frac{\Delta \varepsilon^p}{\varepsilon_f^p} \tag{8}$$

where $\Delta\varepsilon_f^p$ is plastic strain development and ε_f^p is fracture strain, which is defined as follows

$$\varepsilon_f^p = d_1\left(P^* + t^*\right)^{d_2} \tag{9}$$

where d1 and d2 are coefficients, which are determined by the user. The following equation shows the behavior of damaged materials:

$$\sigma_f^* = b\left(P^*\right)^m \left(1 + c\,ln\,\dot{\varepsilon}^*\right) \leq sfma \tag{10}$$

In the above equation, b is the coefficient imposed by the user and *sfma* is the normalized maximum fracture strength. d1 is the controller of amount rate and if it equals to zero, complete degradation occurs in a time interval. In nondamaged materials hydrostatic pressure is as follows:

$$P = k_1\mu + k_2\mu^2 + k_3\mu^3 \qquad (11)$$

where

$$\mu = \frac{\rho}{\rho_0} - 1 \qquad (12)$$

In this equation ρ_0 is the material density and ρ is the material density after deformation. Also according to Hogoniot elastic limit and shear modulus (G), the modulus μ_{HEL} is obtained using the following equation:

$$HEL = k_1\mu_{HEL} + k_2\mu_{HEL}^2 + k_3\mu_{HEL}^3 + (4/3)g(\mu_{HEL}/(1+\mu_{HEL})) \qquad (13)$$

By normalizing the above equation, Eq. (14) is obtained [9]:

$$P_{HEL} = k_1\mu_{HEL} + k_2\mu_{HEL}^2 + k_3\mu_{HEL}^3 \qquad (14)$$

And also

$$\sigma_{HEL} = 1.5\left(HEL - P_{HEL}\right) \qquad (15)$$

Coefficients related to alumina 99/5% are shown in Table 2.4.

TABLE 2.4 Material Coefficients Alumina 99/5% [10]

Equation of State	Linear	Strength	Johnson-Cook
Bulk Modulus A1	2.31×10^8 (KPa)	Shear Modulus	1.52×10^8 (KPa)
Parameter A2	-1.6×10^8 (KPa)	Strain Rate Constant C	0.007 (none)
Parameter A3	2.774×10^9 (KPa)	Intact Strength Constant A	0.88 (none)
Parameter B0	0 (KPa)	Erosion Strain	0.5 (none)

2.3   RESULTS AND DISCUSSION

2.3.1   CERAMIC/ALUMINUM TARGET

Due to penetration of the projectile into the target, ceramic was destroyed as the primary layer, and failure occurred and it caused the loss of projectile head and the velocity reduced due to the hardness and high density. On the other hand, ceramic breaks and goes out of the model in the form of fine particles. As shown in Fig. 2.2 (a), the projectile passes the ceramic in 27 microseconds and changes the curvature in aluminum layer. Then the projectile gets into aluminum layer and completely leave the model by removing the layer in the 53 microseconds, which shows in Fig. 2.2 (b). After complete running of the program, the output speed are achieved and the results are shown in Table 2.5. Comparing the results

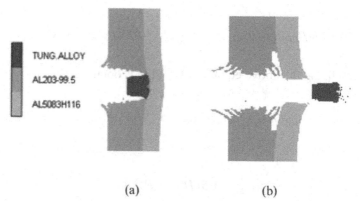

(a)   (b)

FIGURE 2.2   Penetration of projectile at different times: (a) 0.27 microsecond (b) 0.53 microsecond.

TABLE 2.5   Results of Numerical Simulations

Sample	Ceramic thickness (mm)	Aluminum thickness (mm)	Vr (m/s) (sim.)	Vr (m/s) (exp.) [10]	Error (%)
1	20	10	969.73	930	4.27
2	20	15	891.57	930–960	5.65
3	25	10	879.29	960	8.4

of output speed simulations, indicate that by increasing the thickness of ceramic in the third model, the speed is considerably reduced. In contrast, with increasing aluminum thickness and keeping the thickness of the ceramic constant, the output speed is increased. As shown in Fig. 2.3, the slope of diagram at ceramic region decreases gradually by getting into the aluminum layer.

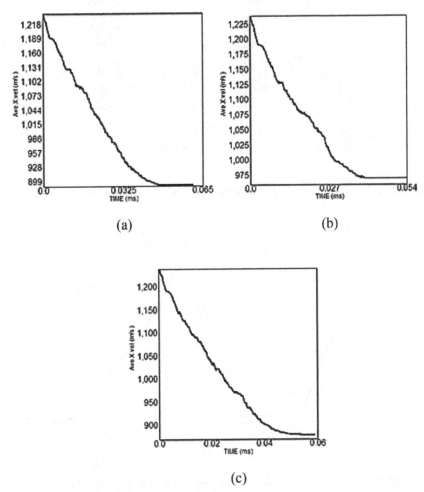

(a)   (b)

(c)

FIGURE 2.3   Output speeds: (a) First sample, (b) Second sample, (c) Third sample.

2.3.2 CERAMIC/ALUMINUM TARGET WITH POLYMER LAYER

After comparing the experimental results with simulation results, now we place the 10-mm layer of Kevlar between the ceramic and aluminum in the first sample and obtain the velocity output (Figs. 2.4 and 2.5). As

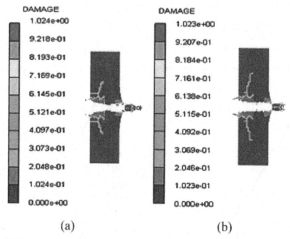

(a)   (b)

FIGURE 2.4   Complete passing of projectile: (a) Kevlar- epoxy, (b) glass-epoxy.

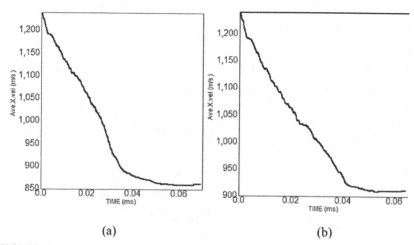

(a)   (b)

FIGURE 2.5   Output speed of tungsten projectile: (a) Kevlar-epoxy, (b) glass-epoxy.

TABLE 2.6   Results of Numerical Simulations

Sample	Ceramic thickness (mm)	Kevlar-epoxy thickness (mm)	Glass-epoxy thickness (mm)	Aluminum thickness (mm)	Vr (m/s) (sim)
1	20	10	–	10	862.75
2	20	–	10	10	911.4

shown in Table 2.6, adding a layer of Kevlar-epoxy will increase ballistic resistance of the panel and the speed of output is reduced by 11%. Then we substitute the layer of glass-epoxy by Kevlar-epoxy and the projectile passes through the model at 911.4 m/s, and the output speed is reduced by 6%. By comparing the results obtained, it can be seen that adding Kevlar-epoxy layer will increase ballistic resistant of panels against projectile penetration.

2.4   CONCLUSION

In this research, the penetration of tungsten projectile into the composite ceramics-metal target is simulated using Autodyne software. Metal ceramic targets are used in armor applications, including coachwork, bulletproof doors, and so on. The behavioral equation used in the simulation is Johnson Cook for projectile and metal and Johnson Holmquist for ceramics. These behavioral equations have different empirical coefficients, by using which the user can obtain more precise results. In each model the output speed of the projectile is calculated and compared with the results of experimental tests. Errors existing in the simulation is less than 10%, which implies a good agreement of the performed simulation and experimental results. Then in order to increase ballistic resistance of the armor, once a layer of Kevlar-epoxy and once again a layer of glass-epoxy was placed between ceramic and aluminum and the obtained output speeds were compared with each other. The simulation results indicate good ballistic resistance of Kevlar-epoxy to the penetration of tungsten projectile.

KEYWORDS

- Autodyne software
- ceramics-aluminum target
- output speed
- polynomial
- tungsten projectile

REFERENCES

1. Kazrayan N., Vahedi Kh. Numerical simulation ballistic long rods projectiles penetration in ceramic/metal targets. Engineering science, 17, 15–23 (1385) (in Persian).
2. Rosenberg Z., Dekel E., Yeshurun Y., Bar-On E. Experiment and 2-D simulations of high velocity penetrations into ceramic tiles. International Journal of Impact Engineering, 17, 697–706 (1995).
3. Sánchez Gálvez V., Sánchez Paradela L. Analysis of failure of add-on armor for vehicle protection against ballistic impact, Engineering Failure Analysis, 16, 1837–1845 (2009).
4. Prakash A., Rajasankar J., Anandavalli N., Verma M., Iyer N. R. Influence of adhesive thickness on high velocity impact performance of ceramic/metal composite targets. International Journal of Adhesion and Adhesives, 41, 186–197 (2013).
5. Madhu V., Ramanjaneyulu K., Balakrishna Bhat T., Gupta N. K. An experimental study of penetration resistance of ceramic armor subjected to projectile impact. International Journal of Impact Engineering, 32, 337–350 (2005).
6. Ning J., Ren H., Gu T., Li, P. Dynamic response of alumina ceramics impacted by long tungsten projectile. International Journal of Impact Engineering, 62, 60–74 (2013).
7. Tan Z.H., Han X., Zhang W., Luo S.H. An investigation on failure mechanisms of ceramic/metal armor subjected to the impact of tungsten projectile. International Journal of Impact Engineering, 37, 1162–1169 (2010).
8. Saedi Daryan A., Jalili S. Blast and Impact Engineering with AUTODYNE applications. Tehran publisher, Darian engineers Publications, (1391) (in Persian).
9. Shafiei M., Vahedi Kh. Analytical and numerical study of long rod projectile penetration into concrete and concrete – steel targets., Journal of Energetic Materials, 8, 43–51 (1392) (in Persian).
10. Chi R., Serjouei A., Sridhar I., Tan, G. Ballistic impact on bi-layer alumina/aluminum armor: A semianalytical approach. International Journal of Impact Engineering, 52, 37–46 (2013).

CHAPTER 3

CHEMICAL MODIFICATION OF SYNDIOTACTIC 1,2-POLYBUTADIENE

M. I. ABDULLIN[1], A. B. GLAZYRIN[1], O. S. KUKOVINETS[1], A. A. BASYROV[1], and G. E. ZAIKOV[2]

[1]Bashkir State University, Ufa, 450077, Russia,
E-mail: ProfAMI@yandex.ru

[2]N.M.Emanuel Institute of Biochemical Physics, Russian Academy of Sciences, 4 Kosygin str., Moscow 119334, Russia,
E-mail: chembio@sky.chph.ras.ru

CONTENTS

ABSTRACT

Results of chemical modification of syndiotactic 1,2-polybutadiene under various chemical reagents are presented. Influence of the reagent nature both on the reactivity of >C=C< double bonds in syndiotactic 1,2-polybutadiene macromolecules at its chemical modification and on the composition of the modified polymer products is considered basing on the analysis of literature and the authors' own researches.

3.1 INTRODUCTION

Obtaining polymer materials of novel or improved properties is considered a major direction in modern macromolecular chemistry [1]. Researches aimed at obtaining polymers via chemical modification methods have been quite prominent along with traditional synthesis methods of new polymer products by polymerization or polycondensation of monomers.

Polymers with unsaturated macrochains hold much promise for modification. The activity of carbon-carbon double bonds as related to many reagents allows introducing substituents of different chemical nature in the polymer chain (heteroatoms including). Such modification helps to vary within wide limits physical and chemical properties of the polymer and render it new and useful features.

Syndiotactic 1,2-polybutadiene obtained by stereospecific butadiene polymerization in complex catalyst solutions [2–7] provides much interest for chemical modification.

In contrast to 1,4-polybutadiens and 1,2-polybutadiens of the atactic structure, the syndiotactic 1,2-polybutadiene exhibit thermoplastic properties combining elasticity of vulcanized rubber and ability to move to the viscous state at high temperatures and be processed like thermoplastic polymers [8–11].

The presence of unsaturated >C=C< bonds in the syndiotactic 1,2-PB macromolecules creates prerequisites for including this polymer into various chemical reactions resulting in new polymer products. Unlike 1,4-polybutadiens, the chemical modification of syndiotactic 1,2-PB is insufficiently studied, though there are some data available [12–15].

A peculiarity of the syndiotactic 1,2-PB produced nowadays is the presence of statistically distributed cis- and trans-units of 1,4-diene polymerization [16, 17] in macromolecules along with the order of 1,2- units at polimerization of butadiene −1,3. Their content amounts to 10–16%. Thus, by its chemical structure, syndiotactic 1,2-polybutadiene can be considered as a copolymer product containing an orderly arrangement of 1,2-units and statistically distributed 1,4 polymerization units of butadiene-1,3:

$$\cdots-CH_2-CH-CH_2-CH-CH_2-CH-CH_2-CH=CH-CH_2-CH-\cdots$$

Taking into account syndiotactic 1,2-PB microstructures and the presence of $>C=C<$ various bonds in the polydiene macromolecules, the influence of some factors on the polymer chemical transformations has been of interest. The factors in question are determined both by the double bond nature in the polymer and the nature of the substituent in the macromolecules.

In the chapter the interaction of the syndiotactic 1,2-PB and the reagents of different chemical nature as ozone, peroxy compounds, halogens, carbenes, aromatic amines and maleic anhydride are considered.

A syndiotactic 1,2-PB with the molecular weight $M_n = (53–72)\times10^3$; $M_w/M_n = 1.8–2.2$; 84–86% of 1,2 butadiene units (the rest being 1,4-polimerization units); syndiotacticity degree 53–86% and crystallinity degree 14–22% was used for modification.

3.2 OZONATION

At interaction of syndiotactic 1,2-PB and ozone the influence of the inductive effect of the alkyl substituents at the carbon-carbon double bond on the reactivity of double bonds in 1,2 and 1,4-units of butadiene polymerization is vividly displayed. Ozone first attacks the most electron-saturated inner double bond of the polymer chain. The process is accompanied by the break in the $>C=C<$ bonds of the main chain of macromolecules and a noticeable decrease in the intrinsic viscosity and molecular weight of the polymer (Fig. 3.1) at the initial stage of the reaction (functionalization degree $\alpha < 10\%$) [17–20]. Due to the ozone high reactivity, partial splitting of the vinyl groups is accompanied by spending double bonds in the main polydiene chain.

However, it does not affect the average molecular weight of the polymer up to the functionalization degree of 15% (Fig. 3.1). Depending on the chemical nature of the reagent used for the decomposition of the syndiotactic 1,2-PB ozonolysis products (dimethyl sulfide or lithium aluminum hydride) [17–20], the polymer products containing aldehyde or hydroxyl groups are obtained (Scheme 1).

The structure of the modified polymers is set using IR and NMR spectroscopy [17]. The presence of C-atom characteristic signals connected with aldehyde (201.0–201.5 ppm) or hydroxyl (56.0–65.5 ppm) groups in ^{13}C NMR spectra allows identifying the reaction products.

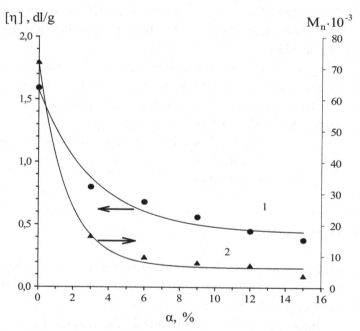

FIGURE 3.1   Dependence of the intrinsic viscosity [η] (*1*) and the average molecular weight M_n (*2*) of the formyl derivative of the syndiotactic 1,2-PB from the functionalization degree of the α polymer (with chloroform as a solvent, 25°C).

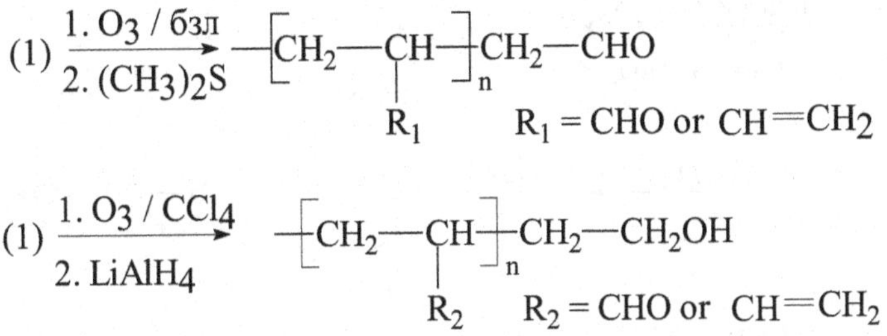

(1) $\dfrac{1.\ O_3\ /\ \text{бзл}}{2.\ (CH_3)_2S}$ $-\!\!\left[CH_2\!-\!\underset{\underset{R_1}{|}}{CH}\right]_n\!\!CH_2\!-\!CHO$

$R_1 = CHO$ or $CH\!=\!CH_2$

(1) $\dfrac{1.\ O_3\ /\ CCl_4}{2.\ LiAlH_4}$ $-\!\!\left[CH_2\!-\!\underset{\underset{R_2}{|}}{CH}\right]_n\!\!CH_2\!-\!CH_2OH$

$R_2 = CHO$ or $CH\!=\!CH_2$

SCHEME 1

Thus, the syndiotactic 1,2-PB derivatives with oxygen-contained groupings in the macromolecules may be obtained via ozonation. Modified 1,2-PB with different molecular weight and functionalization degrees containing hydroxi- or carbonyl groups are possible to obtain by regulating the ozonation degree and varying the reagent nature used for decomposing the products of the polymer ozonation.

3.3 EPOXIDATION

Influence of the double bond polymer nature in the reaction direction and the polymer modification degree is vividly revealed in the epoxidation reaction of the syndiotactic 1,2-PB, which is carried out under peracids (performic, peracetic, meta-chloroperbenzoic, trifluoroperacetic ones) [21–27], tert-butyl hydroperoxide [21–23] and other reagents [28–30] (Table 3.1).

Depending on the nature of the epoxidated agent and conditions of the reaction [21–27], polymer products of different composition and functionalization degree can be obtained (Table 3.1).

As established earlier [21–24, 26], at interaction of syndiotactic 1,2-PB and aliphatic peracids obtained in situ under hydroxyperoxide on the corresponding acid, the $>C\!=\!C<$ double bonds in 1,4-units of macromolecules are mainly subjected to epoxidation (Scheme 2). Higher activity of the double bonds of 1,4-polymerization units in the epoxidation reaction is revealed at interacting syndiotactic 1,2-PB and sodium hypochlorite as well as percarbonic acid salts obtained in situ through the appropriate carbonate and hydroperoxide [31–35].

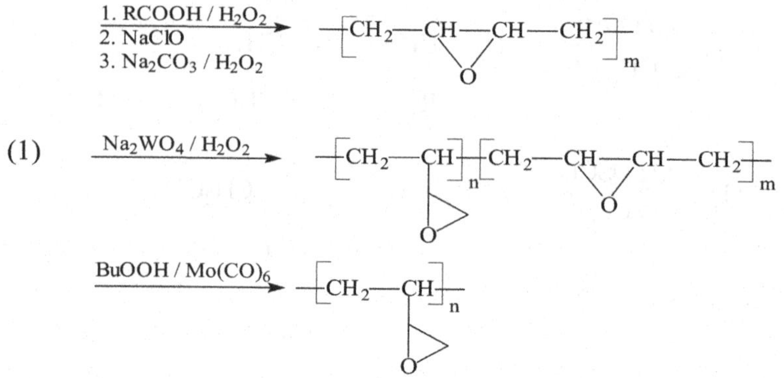

SCHEME 2

It should be noted that the syndiotactic 1,2-PB epoxidation by the sodium hypochlorite and percarbonic acid salts carried out in the alkaline enables to prevent the disclosure reactions of the epoxy groups and a gelation process of the reaction mass observed at polydiene epoxidation by aliphatic peracids [31, 32]. The functionalization degree of 1,2-PB in the reactions with the stated epoxidizing agents ($\alpha \leq 16\%$, Table 3.1) is determined by the content of inner double bonds in the polymer.

TABLE 3.1 Influence of the Epoxidizing Agent on the Functionalization Degree α of Syndiotactic 1,2-PB and the Composition of the Modified Polymer*

| Epoxidizing agent | α, mol. % | Content in the modified polymer, mol.% | | | |
| | | Epoxy groups | | >C=C<bonds | |
		1,2-units	1,4- units	1,2- units	1,4- units
**R^1COOH/H_2O_2	11.0–16.0	-	11.0–16.0	84.0	0–5.0
**R^2COOOH	32.1	16.1	16.0	67.9	-
**R^3COOOH	34.6	18.6	16.0	65.4	-
Na_2WO_4/H_2O_2	31.0	15.0	16.0	69.0	-
Na_2MoO_4/H_2O_2	23.7	7.7	16.0	76.3	-
$Mo(CO)_6/t$-BuOOH	18.0	18.0	-	66.0	16.0
NaClO	16.0	-	16.0	84.0	-
$NaHCO_3/H_2O_2$	16.0	-	16.0	84.0	-

* the polymer with the content of 1,2- and 1,4-units with 84 and 16%, respectively;

** where R^1 – H-, Me-, Et-, $CH_3CH(OH)$-; R^2 – м-ClC_6H_4-; R^3 – CF_3-.

To obtain the syndiotactic 1,2-PB modifiers of a higher degree of functionalization (up to 35%) containing oxirane groups both in the main and side chain of macromolecules (Scheme 2) it is necessary to use only active epoxidizing agents meta-chloroperbenzoic (MCPBA), trifluoroperacetic acids (TFPA) [23], and metal complexes of molybdenum and tungsten, obtained by reacting the corresponding salts with hydroperoxide) [25] (Table 3.1). From the epoxidizing agents given a trifluoroperacetic acid is most active (Fig. 3.2) [23, 26].

However, at the syndiotactic 1,2-PB epoxidation by the trifluoroperacetic acid, a number of special conditions is required to prevent the gelation of the reaction mass, namely usage of the base (Na_2HPO_4, Na_2CO_3 et al.) and low temperature (less than 5°C) [23].

At reacting the syndiotactic 1,2-PB and the catalyst complex [t-BuOOH – Mo(CO)$_6$] a steric control at approaching the reagents to the double polymer bond is carried out [21–24]. This results in participation of less active but more available vinyl groups of macromolecules in the reaction (Table 3.1, Scheme 2).

Thus, modified polymer products with different functionalization degrees (up to 35%) may be obtained on the syndiotactic 1,2-PB basis according to the epoxidizing agent nature. The products in question contain oxirane groups in the main chain (with aliphatic peracids, percarbonic acid

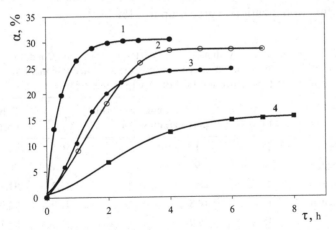

FIGURE 3.2 Influence of the peracid nature on the kinetics of oxirane groups accumulation at the syndiotactic 1,2-PB epoxidation: _1_ – TFPA ([Na_2HPO_4]/[TFPA] = 2; 0°C); _2_ – MCPBA (50°C); _3_ – $Na_2MoO_4H_2O_2$ (55°C); _4_ – $HCOOH/H_2O_2$ (50°C).

salts, and NaClO as epoxidizing agents), in the side units of macromolecules [*t*-BuOOH – Mo(CO)$_6$] or in 1,2- and 1,4-units (TFPA, MCPBA, metal complexes of molybdenum and tungsten).

3.4 HYDROCHLORINATION

In adding hydrogen halides and halogens to the $>C=C<$ double bond of 1,2-PB, the functionalization degree of the polymer is mostly determined by the reactivity of the electrophilic agent. Relatively low degree of polydiene hydrochlorination (10–15%) at interaction of HCl and syndiotactic 1,2-PB [16, 36, 37] is caused by insufficient reactivity of hydrogen chloride in the electrophilic addition reaction by the double bond (Table 3.2). Due to this, more electron-saturated $>C=C<$ bonds in 1,4-units of butadiene polymerization are subjected to modification.

The process is intensified at polydiene hydrochlorination under AlCl$_3$ due to a harder electrophile H$^+$[AlCl$_4$]$^-$ formation at its interaction with HCl [37]. In this case double bonds of both 1,2- and 1,4-polidiene units take part in the reaction (Scheme 3):

Usage of the catalyst AlCl$_3$ and a polar solvent medium (dichloroethane) allows speeding up the hydrochlorination process (Table 3.2) and obtaining polymer products with chlorine contained up to 28 mass.% and the functionalization degree α up to 71% [37].

TABLE 3.2 Influence of the Syndiotactic 1,2-PB Hydrochlorination Conditions on the Functionalization Degree α and Chlorine Content in the Modified Polymer (20–25°C; HCl consumption – 0,2 mol/h per mol 1,2-PB; [AlCl$_3$]=5 mass. %)

Solvent	Reaction time, h	Chlorine content in the polymer, mass. %	α, %	Polymer output, %
Chloroform*	24	4.1	10.5	92.0
Dichloroethane*	24	5.9	15.1	90.3
Chloroform	24	12.6	32.2	91.1
dichloroethane	14	18.8	47.9	92.9
Dichloroethane	18	25.9	66.3	95.1
Dichloroethane	24	27.9	71.2	96.7

*w/o catalyst.

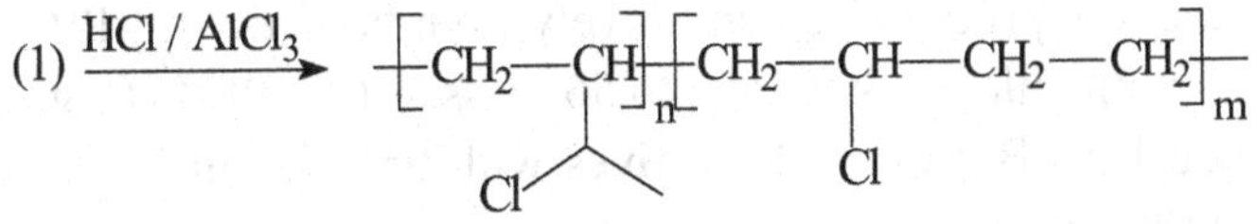

(1) $\xrightarrow{\text{HCl / AlCl}_3}$ $\left[\begin{array}{c} CH_2-CH \\ | \\ Cl \end{array}\right]_n\left[\begin{array}{c} CH_2-CH-CH_2-CH_2 \\ | \\ Cl \end{array}\right]_m$

SCHEME 3

By the ^{13}C NMR spectroscopy method [37] it is established that the $>C=C<$ double bond in the main chain of macromolecules is more active at 1,2 PB catalytic hydrochlorination. Its interaction with HCl results in the formation of the structure (a) (Scheme 4). At hydrochlorination of double bonds in the side chain the chlorine atom addition is controlled by formation of the most stable carbocation at the intermediate stage. This results in the structure (b) with the chlorine atom at carbon β-atom of the vinyl group (Scheme 4):

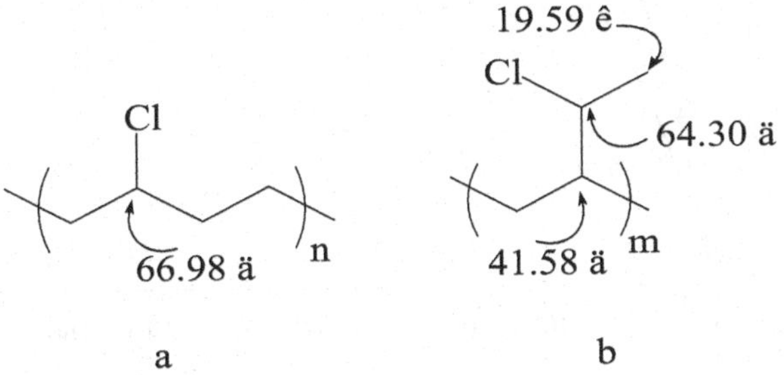

19.59 ê

Cl   64.30 ä

Cl

66.98 ä   n   41.58 ä   m

a   b

SCHEME 4

3.5 HALOGENATION

Effective electrophilic agents like chlorine and bromine easily join double carbon-carbon bonds [16, 38–40] both in the main chain of syndiotactic 1,2-PB and in the side chains of macromolecules (Scheme 5):

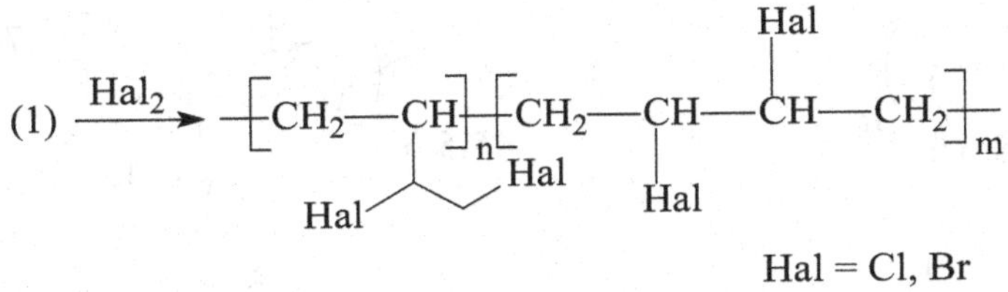

(1) $\xrightarrow{\text{Hal}_2}$ $\left[\begin{array}{c} CH_2-CH \\ | \\ CH_2 \\ | \\ Hal \end{array}\right]_n\left[\begin{array}{c} CH_2-CH-CH-CH_2 \\ | \quad | \\ Hal \quad Hal \\ \\ Hal \end{array}\right]_m$

Hal = Cl, Br

SCHEME 5

The reaction proceeds quantitatively: syndiotactic 1,2-PB chlorine derivatives with chlorine w(Cl) up to 56 mass.% ($\alpha \sim 98\%$) (Fig. 3.3) and syndiotactic 1,2-PB bromo- derivatives with bromine up to 70 mass. % ($\alpha \sim 94\%$) are obtained.

According to the ^{13}C NMR spectroscopy, polymer molecules with dihalogen structural units (Scheme 6) and their statistic distribution in the macro chain serve as the main products of syndiotactic 1,2-PB halogenation [16, 40].

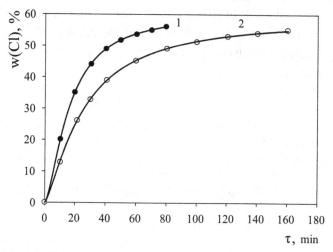

FIGURE 3.3   Kinetics of syndiotactic 1,2-PB chlorination. Chlorine consumption: $1 - 1$ mol/h per mol of syndiotactic 1,2-PB; $2 - 2$ mol/h per mol of syndiotactic 1,2-PB; 20°C, with CHCl3 as a solvent.

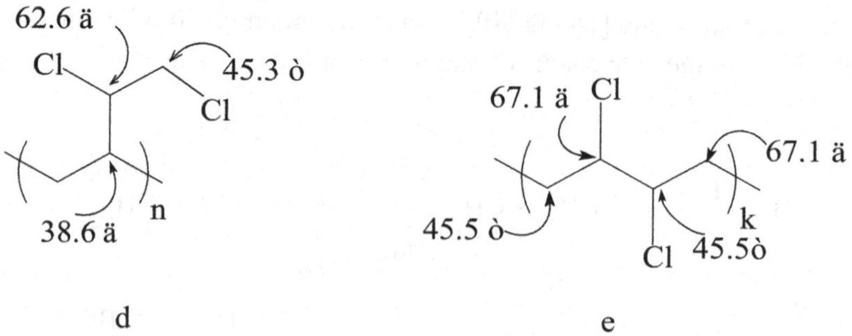

SCHEME 6

3.6 DICHLOROCYCLOPROPANATION

There is an alternative method for introducing chlorine atoms into the 1,2-PB macromolecule structure, namely a dichlorocyclopropanation reaction. It is based on generating an active electrophile agent in the reaction mass at dichlorocarbene modification, which is able to inter-act with double carbon-carbon polymer bonds [41–43]. The syndiotactic 1,2-PB dichlorocyclopropanation is quite effective at dichlorocarbene generating by Macoshi by the chloroform reacting with an aqueous solu-tion of an alkali metal hydroxide. The reaction is carried out at the pres-ence of a phase transfer catalyst and dichlorocarbene addition in situ to the double polydiene links [44–46] according to Scheme 7.

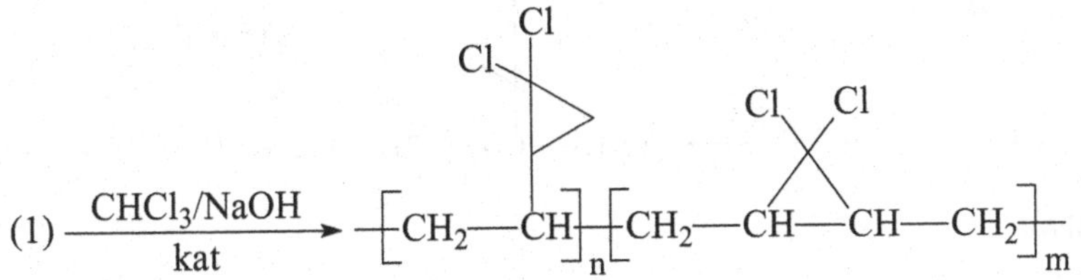

SCHEME 7

The ^{13}C NMR spectroscopy results testify the double bonds dichlo-rocyclopropanation both in the main chain and side chains of polydiene macromolecules (Scheme 8).

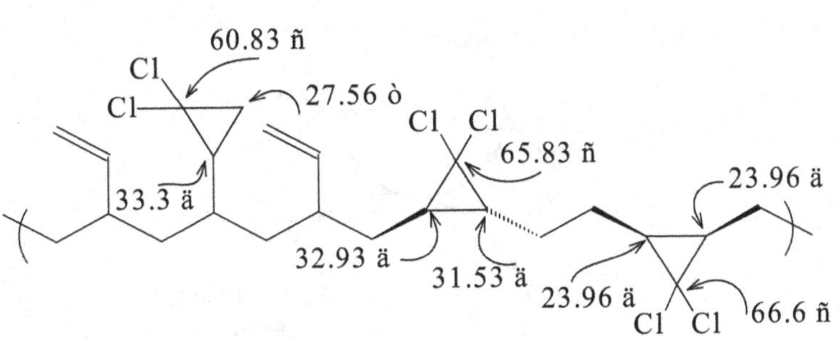

SCHEME 8

Cis-and trans-double bonds in the 1,4-addition units [45] are more active in the dichlorocarbene reaction.

The polymer products obtained contain chlorine up to 50 mass.% which corresponds to the syndiotactic 1,2-PB functionalization degree ~97%, i.e. in the reaction by the Makoshi method full dichlorocarbenation of unsaturated $>C=C<$ polydiene bonds is achieved [23, 45].

3.7   INTERACTION WITH METHYLDIAZOACETATE

Modified polymers with methoxycarbonyl substituted cyclopropane groups [47–51] are obtained by interaction of syndiotactic 1,2-PB and carb methoxycarbonyl generated at a catalytic methyldiazoacetate decomposition in the organic solvent medium (Scheme 9).

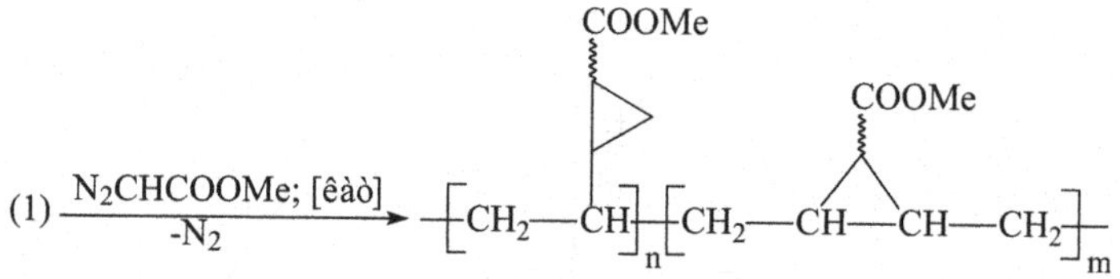

SCHEME 9

The catalytic decomposition of alkyldiazoacetates comprises the formation of the intermediate complex of alkyldiazoacetate and the catalyst [51, 52]. The generated alcoxicarbonylcarbene at further nitrogen splitting is stabilized by the catalyst with the carbine complex formation [51, 52], interaction of which with the alkene results in the cyclopropanation products (Scheme 10):

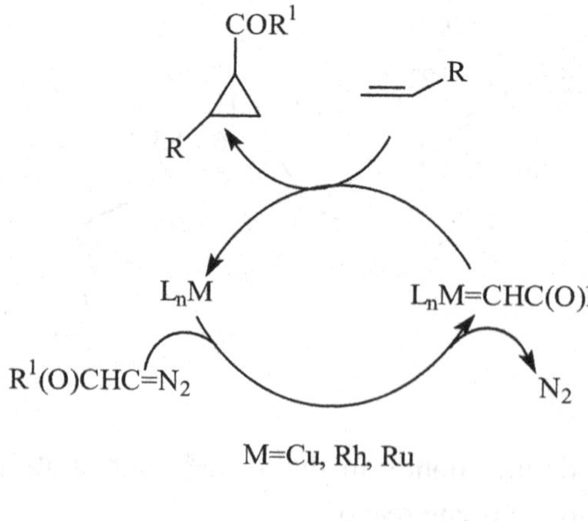

M=Cu, Rh, Ru

SCHEME 10

The output of the cyclopropanation products is determined by the reactivity of the $>C=C<$ double bond in the alkene as well as the stability and reactivity of the carbine complex $L_nM=CH(O)R^1$ which fully depend on the catalyst used [52].

The catalysts applied in the syndiotactic 1,2-PB cyclopropanation range as follows: $Rh_2(OAc)_4$ ($\alpha=38\%$) > [Cu OTf]·0,5 C_6H_6 ($\alpha=28\%$) > $Cu(OTf)_2$ ($\alpha=22\%$) [51].

By the ^{13}C NMR spectroscopy methods it is established that in the presence of copper (I), (II) compounds, double bonds both of the main chain and in the side units of syndiotactic 1,2-PB macromolecules are subjected to cyclopropanation whereas at rhodium acetate $Rh_2(OAc)_4$ mostly the $>C=C<$ bonds in the 1,4- addition units undergo it [51].

Thus, catalytic cyclopropanation of syndiotactic 1,2-PB under methyldiazoacetate allows obtaining polymer products with the functionalization degree up to 38% and their macromolecules containing cyclopropane groups with an ester substituent. The determining factor influencing the cyclopropanation direction and the syndiotactic 1,2-PB functionalization degree is the catalyst nature. By using catalysts of different chemical nature it is possible to purposefully obtain the syndiotactic 1,2-PB derivatives containing cyclopropane groups in the main chain (with rhodium acetate as a catalyst) or in 1,2- and 1,4-polydiene units (copper compounds) respectively.

Along with the electronic factors determined by different electron saturation of the $>C=C<$ bonds in 1,2- and 1,4-units of polydiene addition and the catalyst nature used in modification, the steric factors may also influence the reaction and the syndiotactic 1,2-PB functionalization degree. The examples of the steric control may serve the polydiene reactions with aromatic amines and maleic anhydride apart from the above considered epoxidation reactions of syndiotactic 1,2-PB by *tret*-butyl hydroxyperoxide.

3.8 INTERACTION WITH AROMATIC AMINES

Steric difficulties prevent the interaction of double bonds of the main chain of syndiotactic 1,2-PB macromolecules and aromatic amines (aniline, N, N-dimethylaniline and acetanilide). In the reaction with amines

[17, 20, 23] catalyzed by $Na[AlCl_4]$ the vinyl groups of the polymer enter the reaction and form the corresponding syndiotactic 1,2-PB arylamino derivatives (Scheme 11):

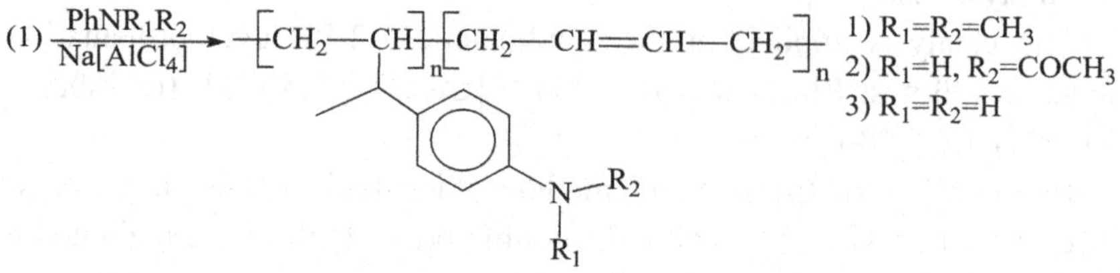

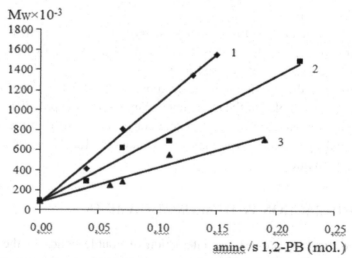

SCHEME 11

From the NMR spectra analysis it is seen that the polymer functionalization is held through the β-atom of carbon vinyl groups [17].

Introduction of arylamino groups in the syndiotactic 1,2-PB macromolecules leads to increasing the molecular weight M_w (Fig. 3.4) and the size of macromolecular coils characterized by the mean-square radius of gyration $(\overline{R^2})^{1/2}$ [17, 20].

The results obtained indicate to the intramolecular interaction of monomer units modified by aromatic amines with vinyl groups of polydiene macromolecules at syndiotactic 1,2-PB modification. This leads to the

FIGURE. 3.4   Influence of the aromatic amine nature on the molecular weight (M_w) of the polymer modified by: _1_ – acetanilide; _2_ – N,N-dimethylaniline; _3_ – aniline.

formation of macromolecules of the branched and linear structure (Scheme 12):

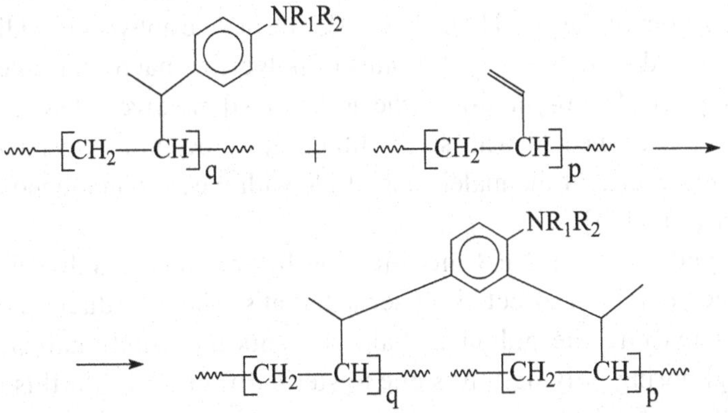

SCHEME 12

Steric difficulties determined by the introduction of bulky substituents in the polydiene units ("a neighbor effect" [53, 54]) does not allow to obtain polymer products with high functionalization degree as the arylamino groups in the modified polymer does not exceed 8 mol.%. At the same time secondary intermolecular reactions are induced in the synthesis process involving arylamino groups of the modified macromolecules and result in the formation of linear or branched polymer products with high molecular weight.

3.9   INTERACTION WITH MALEIC ANHYDRIDE

The polymer products with anhydride groups are synthesized by thermal adding (190°C) of the maleic anhydride to the syndiotactic 1,2-PB [23, 24] (Scheme 13):

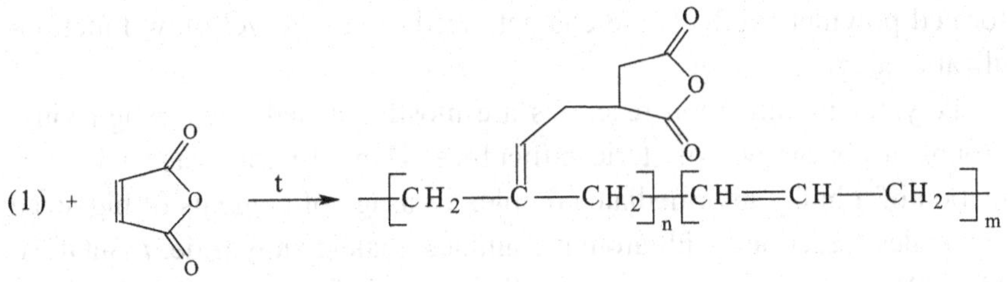

SCHEME 13

The ^{13}C NMR spectroscopy results show that the maleic anhydride addition is carried out as an ene-reaction [55] by the vinyl bonds of the polymer without the cycle disclosure and the double bond is moved to the β-carbon atom of the vinyl bond [23, 24]. The maleic anhydride addition to the >C=C< double bonds of 1,4-units of polydiene macromolecules does not take place. As in synthesis of the arylamino derivatives of syndiotactic 1,2-PB, it is connected with steric difficulties preventing the interaction of bulk molecules of the maleic anhydride with inner double bonds of the polymer chain [23].

At syndiotactic 1,2-PB modification by the maleic anhydride, the so-called "neighbor effect" is observed, that si, the introduction of bulk substituents into the polymer chain prevents the functionalization of the neighboring polymer units due to steric difficulties. For this reason the content of anhydride groups in the modified polymer molecules do not exceed ~15 mol. %.

Thus, >C=C< double bonds in 1,2- and 1,4-units of syndiotactic 1,2-PB macromolecules considerably differ in the reactivity due to the polydiene structure. The inductive effect of the alkyl substituents resulting in the increase of the electron density of the inner double bonds of macromolecules determines their high activity in the considered reactions with different electrophilic agents.

At interaction of syndiotactic 1,2-PB with strong electrophiles (ozone, halogens, dichlorocarbene) both inner double bonds and side vinyl groups of polydiene macromolecules are involved in the reaction. It results in polymer products formation with quite a high functionalization degree. In the case when the used reagent does not display enough activity (interaction of syndiotactic 1,2-PB and hydrogen chloride and aliphatic peracids), the process is controlled by electronic factors: more active double bonds in 1,4-units of the polymer chain are subjected to modification whereas the formed polymer products are characterized by a relatively low functionalization degree.

Polymer modification reactions are mostly carried out through vinyl groups at appearance of steric difficulties. They are connected with formation of a bulky intermediate complex or usage of reagents of big-sized molecules (reactions with aromatic amines, maleic anhydride, t-BuOOH/ Mo(CO)$_6$). Such reactions are controlled by steric factors. The predominant course of the reaction by the side vinyl groups of polymer macromolecules is determined by their more accessibility to the reagent attack. However, in such reactions high functionalization degree of syndiotactic 1,2-PB cannot be achieved due to steric difficulties arousing through the introduction of bulky substituents into the polymer chain. They limit the reagent approaching to the reactive polydiene bonds.

Thus, a targeted chemical modification of the polydiene accompanied by obtaining polymer products of different content and novel properties can be carried out using differences in the reactivity of $>C=C<$ double bonds of syndiotactic 1,2-PB. Various polymer products with a set complex of properties is possible to obtain on the syndiotactic 1,2-PB basis varying the nature of the modifying agent, a functionalization degree of the polymer and synthesis conditions.

KEYWORDS

- chemical modification
- functionalization degree
- polymer products
- syndiotactic 1,2-polybutadiene

REFERENCES

1. Kochnev, A. M., Galibeev, S. S. Modification structures and properties of polymers. *Chemistry and Chemical Technology* (in Rus.). 2003, 46, 4, 3–10.
2. Byrihina, N. N. Aksenov, V. I., Kuznetsov, E. I. Syndiotactic 1,2-polybutadiene synthesis process. Patent RU 2177008. 2001.
3. Ermakova, I. Drozdov, B. D., Gavrilova, L. V., Shmeleva, N. V. Syndiotactic 1,2-polybutadiene synthesis process. Patent RU 2072362. 1998.
4. Luo Steven. Iron-based catalyst composition and process for producing syndiotactic 1,2-poly butadiene. Patent US 6284702. 2002.
5. Wong Tang Hong, Cline James Heber. Syndiotactic 1,2-polybutadiene synthesis. Patent US 5986026. 2000.
6. Ni Shaoru, Zhou Zinan, Tang Xueming. The chain structure of 1,2-polybutadienes prepared with molybdenum catalyst systems. *Chinese Journal of Polymer Sci.*, 1983, 2, 101–107.

7. Monteil, V., Bastero, A., Mecking, S. 1,2-Polybutadiene latices by catalytic polymerization in aqueous emulsion. *Macromolecules*, 2005, *38,* 5393–5399.
8. Obata, Y., Tosaki Ch., Ikeyama, M. Bulk Properties of Syndiotactic 1,2-Polybutadiene *Polym. J.* 1975. *Vol. 7,* № 2, 207 – 216.
9. Glazyrin, A. B., Sheludchenko, A. V., Zaboristov, V. N., Abdullin, M. I. Physical, mechanical and rheological properties of syndiotactic 1,2-polybutadiene. *Plastics* (in Rus.). 2005, 8, 13–15.
10. Abdullin, M. I., Glazyrin, A. B., Sheludchenko, A. V., Samoilov, A. M., Zaboristov, V. N. Viscoelastic and rheological properties of the syndiotactic 1,2-polybutadiene. *Journal of Applied Chemistry* (in Rus.). 2007, 80, № 11, 1913–1917.
11. Xigao, J. Epoxidation of unsaturated polymers with hydrogen peroxide. *Polymer Sci.* 1990. 28, №. 9. 285 – 288.
12. Kimura, S., Shiraishi, N., Yanagisawa, S. A new thermoplastic, 1,2-polybutadiene JSR RB- properties and applications. *Polymer-Plastics Technology and Engineering*, 1975, 5, № 1, 83–105.
13. Lawson, G. Carboxylated syndiotactic 1,2-polybutadiene Patent US 4960834. 2003.
14. Gary, L. Catalytic polymerization. Patent US 5278263. 2001.
15. Dontsov, A. A., Lozovik, G. Y. Chlorinated polymers. Khimiya (Chemistry, in Rus): Moscow. 1979. 232 p.
16. Asfandiyarov, R. N. Synthesis and properties of halogenated 1,2-polybutadienes. Thesis Ph.D. Chem. Science. – Ufa, Bashkir State University, 2008. 135 p.
17. Kayumova, M. A. Synthesis and properties of the oxygen-and aryl-containing derivatives of 1,2-syndiotactic polybutadiene. Thesis Ph.D. Chem. Science. – Ufa, Bashkir State University, 2007. 115 p.
18. Abdullin, M. I., Kukovinets, O. S., Kayumova, M. A., Sigaeva, N. N., Ionova I. A., Musluhov, R. R., Zaboristov, V. N. Formyl derivatives of syndiotactic 1,2-polybutadiene. *Polymer Sci. Ser. B.* 2004. 46, № 10, 1774–1778.
19. Gainullina, T. V., Kayumova, M. A., Kukovinets, O. S., Sigaeva, N. N., Muslukhov, R. R., Zaboristov, V. N., Abdullin, M. I. Modification of syndiotactic 1,2-polybutadiene by epoxy groups. *Polymer Sci. Ser. B.* 2005. 47, № 9, 248–252.
20. Abdullin, M. I., Kukovinets, O. S., Kayumova, M. A., Sigaeva, N. N., Musluhov, R. R. Benzoamino derivatives of syndiotactic 1,2-polybutadiene. *Bashkir Chemistry Journal* (in Rus.). 2006. 13, № 1, 29–30.
21. Gainullina, T. V., Kayumova, M. A., Kukovinets, O. S., Sigaeva, N. N., Muslukhov, R. R., Zaboristov, V. N., Abdullin, M. I. Modification of syndiotactic 1,2-polubutadiene by epoxy groups. *Polymer Sci. Ser. B.* 2005. 47, № 9, 1739–1744.
22. Abdullin, M. I., Gaynullina, T. V., Kukovinets, O. S., Khalimov, A. R., Sigaeva, N. N., Musluhov, R. R., Kayumova, M. A. Synthesis and properties epoxy derivatives of syndiotactic 1,2-polybutadiene. *Journal of Applied Chemistry* (in Rus.). 2006. 79, № 8, 1320–1325.
23. Abdullin, M. I., Glazyrin, A. B., Kukovinets, O. S., Basyrov, A. A. Chemical modification of syndiotactic 1,2-polybutadiene. *Chemistry and Chemical Technology* (in Rus.). 2012. 55, № 5, 71–79.
24. Kukovinets, O. S., Glazyrin, A. B., Basyrov, A. A., Dokichev, V. A., Abdullin, M. I. Advances in chemical modification of syndiotactic 1,2-polybutadiene. *Proceedings of the Ufa Scientific Center, Russian Academy of Sciences* (in Rus.). 2013. 1, 29–37.

25. Valekzhanin, I. V., Abdullin, M. I., Glazyrin, A. B., Kukovinets, O. S., Basyrov, A. A. Methods selective epoxidation syndiotactic 1,2-polybutadiene. Current *Problems humanities and natural Sciences* (in Rus.), 2012. 3, 13–14.

26. Abdullin, M. I., Basyrov, A. A., Kukovinets, O. S., Glazyrin, A. B., Khamidullina, G. I. Epoxidation of syndiotactic 1,2-polybutadiene with peracids. *Polymer Sci., Ser. B.* 2013. 55, № 5–6, 349–354.

27. Abdullin, M. I., Glazyrin, A. B., Kukovinets, O. S., Basyrov, A. A. R heological properties of syndiotactic 1,2-polybutadiene and a modified product thereof. *International research Journal* (in Rus.), 2012. 4, 36–39.

28. Abdullin, M. I., Glazyrin, A. B., Kukovinets, O. S., Valekzhanin, I. V., Klysova, G. U., Basyrov, A. A., Methods for synthesizing epoxidized 1,2-polybutadiene. Patent RU 2465285. 2012.

29. Abdullin, M. I., Glazyrin, A. B., Kukovinets, O. S., Valekzhanin, I. V., Kalimullina, R. A., Basyrov, A. A. Methods for synthesizing epoxidized 1,2-polybutadiene. Patent RU 2456301. 2012.

30. Kurmakova, I. N. Structure formation in solutions of epoxy oligomers. *Polymer Sci. Ser. B.* 1985. 21, №12, 906–910.

31. Hayashi, O., Kurihara, H., Matsumoto, Y. Process for producing hydrophilic polymers. Patent US 4528340. 1985.

32. Blackborow, John, R. Epoxidation polybutylene with peroxigen compound, isomerization to form carbonyl compounds. Patent US 5034471. 1991.

33. Jacobi, M. M., Viga, A., Schuster, R. H. Study of the Epoxidation of Polydiene Rubbers II. *Raw Materials and Application.* 2002. 3, 82–89.

34. Emmons, W. D., Pagano, A. S. Peroxytrifluoroacetic Acid. IV. The Epoxidation of Olefins. *J. Am. Chem. Soc.* 1955. 77, № 1, 89–92.

35. Xigao Jian, Allan, S. Hay. Catalytic epoxidation of styrene–butadiene triblock copolymer with hydrogen peroxide. *J. Polym. Sci.: Polym. Chem. Ed.* 1991. 29, 1183–1189.

36. Abdullin, M. I., Glazyrin, A. B., Asfandiyarov, R. N. Chlorine derivatives of syndiotactic 1,2-poly butadiene. *Polymer Sci. Ser. B.* 2009. 51, №8, 1567–1572

37. Glazyrin, A. B., Abdullin, M. I., Muslukhov, R. R., Kraikin, V. A. Hydrochlorinated derivatives of syndiotactic 1,2-polybutadiene. *Polymer Sci., Ser. A.* 2011. 53, № 2, 110–115.

38. Abdullin, M. I., Glazyrin, A. B., Akhmetova, V. R., Zaboristov, V. N. Chlorine derivatives of syndiotactic 1,2-polybutadiene *Polymer Sci., Ser. B.* 2006. 48, № 4, 104–107.

39. Abdullin, M. I., Glazyrin, A. B., Asfandiyarov, R. N. Chlorinated polymers of low molecular weight 1,2-polybutadiene. *Journal Applied Chemistry* (in Rus.). 2007. 10, 1699–1702.

40. Abdullin, M. I., Glazyrin, A. B., Asfandiyarov, R. N., Akhmetova, V. R. Synthesis and properties of halogenated 1,2-syndiotactic polybutadiene. *Plastics* (in Rus.). 2006. 11, 20–22.

41. Lishanskiy, I.S, Shchitokhtsev, V. A., Vinogradova, N. D. Reactions of carbenes with unsaturated polymers. *Polymer Sci., Ser. B.* 1966. 8, 186–171.

42. Komorski, R. A., Horhe, S. E., Carman, C. J. Carbon-13 NMR microstructural characterization of dichlorocarbene adducts of polybutadiene. *J. Polym. Sci.: Polym. Chem. Ed.* 1983. 21, 89 – 96.

43. Nonetzny, A., Biethan, U. Zur Anlagerung von Dichlorcarben an niedermolekulare cis- und Vinyl-cis-Polybutadiene. *Angew. makromol. Chem.* 1978. 74, 61–79.
44. Glazyrin, A. B., Abdullin, M. I., Kukovinets, O. S. Syndiotactic 1,2-polybutadiene: properties and chemical modification. *Herald of Bashkir University* (in Rus.). 2009. 14, №3, 1133–1140.
45. Glazyrin, A. B., Abdullin, M. I., Muslukhov, R. R. Dichlorocyclopropane derivatives of syndiotactic 1,2-polybutadiene. *Polymer Sci., Ser. B.* 2012. 54, 234–239.
46. Glazyrin, A. B., Abdullin, M. I., Khabirova, D. F., Muslukhov, R. R. Method of producing polymers containing *dichlorocyclopropane groups.* Patent RU 2456303. 2012.
47. Glazyrin, A. B., Abdullin, M. I., Sultanova, R. M., Dokichev, V. A., Muslukhov, R. R., Yangirov, T. A., Khabirova, D. F. Method of producing polymers containing cyclopropane group. Patent RU 2443674. 2012.
48. Glazyrin, A. B., Abdullin, M. I., Sultanova, R. M., Dokichev, V. A., Muslukhov, R. R., Yangirov, T. A., Khabirova, D. F. Method of producing polymers containing cyclopropane group. Patent RU 2447055. 2012.
49. Gareyev, V. F., Yangirov, T. A., Kraykin, V. A., Kuznetsov, S. I., Sultanova, R. M., Biglova, R. Z., Dokichev, V. A. Cu(OAc)(2)-2,4-Lutidine-ZnCl2 as an Effective Catalyst of Functionalization of Isobutylene Oligomers and 1,2-Polybutadiene with Methyl Diazoacetate. *Herald of Bashkir University* (in Rus.). 2009. 14, №1, 36–39.
50. Gareyev, V. F., Yangirov, T. A., Volodina, V. P., Sultanova, R. M., Biglova, R. Z., Dokichev, V. A. Cu(OAc)(2)-2,4-Lutidine-ZnCl2 as an Effective Catalyst of Functionalization of Isobutylene Oligomers and 1,2-Polybutadiene with Methyl Diazoacetate. *Journal Applied Chemistry* (in Rus.). 2009. 83, №7, 1209–1212.
51. Glazyrin, A. B., Abdullin, M. I., Dokichev, V. A., Sultanova, R. M., Muslukhov, R. R., Yangirov, T. A. Synthesis and properties of cyclopropane derivatives of polybutadienes. *Polymer Sci., Ser. B.* 2013. 55, 604–609.
52. Shapiro Ye.A., Dyatkin, A. B., Nefedov, O. M. Diazoether. Nauka (Sciences, in Rus.): Moscow. 1992. 78 p.
53. Fedtke, M. Chemical reactions of polymers. Khimiya (Chemistry, in Rus): Moscow. 1990. 152 p.
54. Kuleznev, V. N., Shershnev, V. A. The chemistry and physics of polymers. KolosS (in Rus.): Moscow. 2007. 367 p.
55. Vatsuro, K. V., Mishchenko, G. L. Named Reactions in Organic Chemistry. Khimiya (Chemistry, in Rus): Moscow. 1976. 528 p.

CHAPTER 4

SPECIFICS OF SELECTIVE WETTING OF SOME METAL SUBSTRATES AND THEIR OXIDES

D. A. NGUYEN,[1] I. A. STAROSTINA,[1] O. V. STOYANOV,[1] and D. V. VEZENOV[2]

[1]*Kazan National Research Technological University, 420015, Kazan, K. Marx str., 68, Russia, E-mail: ov_stoyanov@mail.ru*

[2]*Lehigh University, 27 Memorial Dr. W. Bethlehem, PA 18015, USA*

CONTENTS

ABSTRACT

The surface free energy of some metal substrates and their oxides under selective wetting conditions is determined in order to exclude the influence of atmospheric adsorption. The dependence of the obtained values on the surface tension of used neutral hydrocarbon is detected. The pronounced hydrophobicity of studied surfaces before and after thermooxidation is reported.

4.1 INTRODUCTION

Creation of optimal polymer-metal adhesive compound is complicated first of all by the correct selection of adhesive and adherend. Most often, this problem is solved by the trial and error method, and therefore a development of a uniform scientific approach to the adhesive compound design is very relevant.

The surface energy of connected materials has a great importance in the successful implementation of interfacial interaction, so to solve this task requires knowledge of the free surface and interfacial energy of all adhesive compound materials and all their components (Lifshitz – van der Waals component and the acid-base component (γ^{LW} and γ^{AB}), as well as acid and basic parameters (γ^+ and γ^-)).

For metals, the surface free energy (SFE) defines properties such as hardness, wear resistance, corrosion resistance, and the lower the surface energy of the metal, the worse the mechanical properties of its surface.

At a very high value of the surface energy, a metal can become brittle, which can also lead to a decrease of wear resistance.

For polymers and polymer composites, determination of surface characteristics is traditionally carried out using methods of wetting in air.

As for the high-energy surfaces of metals and metal oxides, it is known that they adsorb water vapor and other impurities from the atmosphere, whereby the measured SFE value is significantly reduced.

There are several methods of measuring the surface energy of solids, of which the most reliable results are obtained by the zero creep method (Tamman-Udine) [1, 2], based on the existence of a viscous creep of the body, the ability to slowly flow under the applied force at sufficiently high temperatures.

The graphic interpolation of this force to value, at which viscous creep is counterbalanced by surface tension, allows the surface energy estimation.

A method of measuring the surface tension of metals in the solid phase includes heating of a thread-like pattern at the local point up to the desired creep temperature by creating the oppositely directed temperature gradient from two sides of this point with subsequent determination of the arising tightening efforts in the sample creep area.

For fragile bodies, for which the method of zero creep is not applicable, the method of splitting according to the cleavage planes developed by Obreimov [3] is used, which allows measuring the force which must be applied in order for the crack formed in advance in a solid can develop further. It is clear that both methods are not suitable for evaluation of the surface energy of real metal adherend.

Adhesive technologies deal generally with the *metal substrate*, a sheet or tube metal rolling, the actual surface of which is always covered by the oxide film, which gives an additional surface roughness. In addition to numerous irregularities of substrate surface due to the crystalline nature of the metal, the grooves and deep scratches formed during the manufacturing process can be found in the sample.

Thus, the surface energy of the metal substrate substantially differs from that of the metal. Therefore, to solve our problems, we cannot use the data, reported in scientific literature about the high surface energy of metals, defined by the above-mentioned traditional methods.

From the all stated the conclusion follows about the necessity of correct evaluation of the SFE components and parameters of metal substrates used in adhesive joints. Of particular scientific interest is the development of methods for the measurement not in air, but in other environment, for example, in neutral hydrocarbon, i.e. under conditions of selective wetting, in order to exclude the effects of atmospheric adsorption.

4.2   EXPERIMENTAL SECTION

4.2.1   MATERIALS

Double-distilled water, formamide, glycerin, dimethyl sulfoxide (DMSO), dimethylformamide (DMF) and 88%solution of phenol in water were used as test liquids. SFE components and parameters of test liquids were taken from our earlier work [5].

St3 steel (content of carbon 0.14 – 0.22%) and copper plates were used as metal substrates. Plates were treated with sandpaper to the purity class 10, purified by carbon tetrachloride and immediately placed in a cell of optical glass with a neutral hydrocarbon (nonoxidized metal substrates).

For obtaining oxide films, the plates prepared as described above, were treated in an oven at 190°C for 2 and 4 h.

The roughness of the oxidized and nonoxidized samples was evaluated by using a *MultiMode* scanning probe microscope (Veeco, USA).

Contact angle measurements were conducted using a KM-8 cathetometer in neutral hydrocarbons with different values of free surface energy: n-hexane (18.43 mN/m), n-hexadecane (27.47 mN/m) [4] and with $\gamma_l = \gamma_l^{LW}$ for each of them.

Droplets of test liquids were applied with a syringe with a needle cut at right angles. Photographing of droplets was carried out in a macro mode.

4.3   RESULTS AND DISCUSSION

The phenomenon of selective wetting in relation to the metals was considered in the works of P.A. Rebinder, B.D. Summ and S.S. Voyutsky [6–8].

Currently, it is used mostly for investigation and prevention of metal corrosion near the interface between two immiscible phases: an aqueous electrolyte – hydrocarbon [9, 10].

The most common case of selective wetting is the contact of a solid with a polar liquid (water, aqueous solutions) and nonpolar hydrocarbon liquid (conventionally called oil). These systems are of great practical importance (for example, in the technology of flotation processes). It is known that the selective wetting of a solid surface by nonpolar hydrocarbon is observed when the difference of the polarities between hydrocarbon and a solid surface is smaller than that between water and this surface.

Of special interest is the study of the surface properties of metal substrates covered by oxide film. In recent works, we evaluated the surface-energy characteristics of a variety of metal surfaces under selective wetting conditions [11, 12]. It was found that the values of the SFE components and parameters depend on the hydrocarbon used – the smaller its surface tension, the higher the obtained total SFE value of metal surface.

In addition, it was noted that the increase of surface hydrophilicity due to oxidation takes place to a very small extent (or not observed at all). In the scientific literature, there is a significant number of reports concerning the hydrophilicity of oxidized metals. Beginning from P.A. Rebinder's works on physical chemistry of flotation processes [6] till the present time, this statement is repeated in numerous publications devoted to the study of the corrosion of oil-field equipment [9,10].

Summ [7] and Voyutsky [8] consider metal oxides and hydroxides as hydrophilic, along with quartz, glass, calcite, silicates, carbonates, sulfates and halides of alkali and alkaline earth metals, as these are substances with a strong intermolecular interaction. Therefore, the more detailed study of the wettability and surface-energy properties of oxidized metal substrates was of interest.

We studied the surfaces of thermally treated and untreated metal substrates by atomic force microscopy (AFM). The photographs and profiles of the St3 steel surfaces are presented below:

Large irregularities (a 2 μm in width and larger scratches) were found on the surface of thermally treated steel sample. The maximum roughness value on a 1×1 μm site (no large irregularities) was 100 nm.

For the thermally untreated steel, the maximum roughness value on a 1×1 μm site (no large irregularities) was 60 nm. Thus, the thermally treated samples show greater roughness on smooth surface sites. It is obvious that the oxide film formed on metal surfaces has a highly developed rough surface.

We have found that thermal treatment of steel and especially copper leads to the decrease of their wetting with water in the presence of hexane and hexadecane.

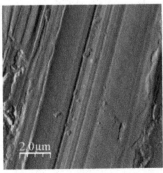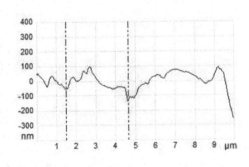

FIGURE 4.1   AFM image of the thermally untreated steel surface and its cross-section.

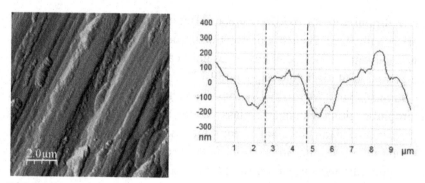

FIGURE 4.2 AFM image of the thermally treated steel surface and its cross-section.

For all surfaces under study, the oxidation was carried out in 2 modes – for 2 and 4 h at 190°C.

It was found that under more "strong" thermal treatment the hydrophobic properties of metal substrates increase (Fig. 2, b) but for copper treated for 4 h at 190°C, a complete nonwettability is observed (Fig. 2, d).

The results obtained evidenced the hydrophobicity of not only metals, but, even to a greater extent, their oxides, and thus disagreed with the literature data available. However, this contradiction, may have a very simple explanation.

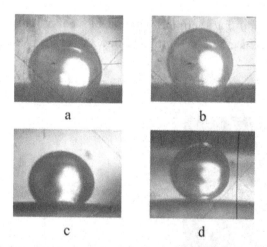

a

b

c

d

FIGURE 4.2 A drop of water on the metal substrate in the presence of hexadecane; a – nonoxidized steel, b – steel oxidized for 4 h at 190°C, c – nonoxidized copper, d – copper oxidized for 4 h at 190°C.

According to MacCafferty [13], the oxide film surface on the metal substrate always has an external layer that contains the hydroxyl groups.

These hydroxyl groups may have either acidic or basic nature, depending on the metal cation present in the oxide film. Therefore, oxides formed on the surface of the substrates during the production and storage may also have both acidic and basic (Lewis) nature.

If this substrate is placed in a neutral hydrocarbon, and the droplets of water solutions with different pH values are then applied on the surface, the observed wetting angles will cross a maximum at the isoelectric point of the surface oxide.

An isoelectric point (IEP) is the characteristic of the status of the disperse phase particle surface, at which the electrokinetic potential is equal to zero. At IEP the electrokinetic phenomena are not observed. In disperse systems the electric charge of particle surface is caused by either partial dissociation of surface ionogenic groups or the adsorption of potential-determining ions from solution. Near the charged surface, a double electric layer is formed, with the potential-determining ions being located at one side of which near the interface, and the counter ions – at another, external side. Depending on the concentration of the potential-determining ions and the specifically adsorbed counter ions, the electrokinetic potential value may vary from positive to negative, being equal to zero at the IEP. In the absence of specific adsorption of counter ions the IEP coincides with the potential of zero charge of the surface.

If the pH-value of the test solution is equal to the IEP of oxide, the hydroxyl groups may remain undissociated. In this case, the angle of wetting of a drop has the maximum value. If the pH-value of test solution is above or below the IEP, the surface interacts with the H^+ or OH^- ions and becomes positively or negatively charged. In this case, the contact angle of wetting of drops decreases.

Earlier we studied the isoelectric point shifts due to thermal oxidation for a number of metal substrates of various steel grades, titanium and brass, and found that the IEP of each metal substrate varies due to thermal oxidation of the surface, that seems as quite logical [12, 14]. For brass, for example, the IEP shift is a result of oxidation from 8 down to 6.

It is logical to assume that the pH-value of test water is close to and, in the case of oxidized copper, coincides with the IEP of the surfaces of metal substrates under study. In pure water the equilibrium with atmospheric carbon dioxide is established quickly (within 10–20 min), with the pH-value being 5.5–6, which indicates the presence of a large number of hydrogen ions, approximately 900 times more than the OH⁻ions. This may explain mainly acidic nature of the normal clean water [15]. The influence of the pH-value on wetting was also studied in the work of B.D. Summ, who noted that the pH-value may influence the nature of the solid surface and change the conditions of adsorption [7]. According to many scientific reports on hydrophilic properties of metal oxides, water contains dissolved gases and impurities, which leads to a significant shift of the pH-value from the IEP value of surface [9, 10, 16] resulting in the improved wettability.

Thus, an unambiguous statement about the hydrophilicity of metal oxide surfaces is wrong. Both a nonpolar hydrocarbon and test liquid can wet the surface under study, so in each case a competition between them will take place. The form of a drop will be determined by the ratio of intermolecular interactions between individual liquids as well as liquids with a solid. A great selective wetting in relation to this surface will be shown by the fluid the polarity value of which is closer to that of the solid body, and by spreading of which the surface energy of the system will decrease by a greater value [6, 8]. It is worthwhile noting that the polarity of the copper surface is not high compared with a large number of metal substrates used in adhesive technologies. As was shown in our previous studies, the SFE of copper substrate, as determined in the air, is 32.0 mN/m, of which 7.2 mN/m refers to the acid-base component, which are less than the same values for brass, aluminum and steels of different grades [12]. The surface tension value of hexadecane is 27.5 mN/m, consequently, the SFE values of copper and hexadecane are close to each other.

The contact angles of wetting metal substrates with test liquids in hexadecane as well as the work of adhesion are shown in Table 4.1.

The obtained results demonstrate the oleophilicity of the surfaces under study irrespective of the degree of oxidation. The contact angle cosine of the wetting with water in all cases is negative. Both oxidized and unoxidized copper also have the pronounced liophobic ability. Consequently, a great selective wetting in relation to the metal surfaces studied is shown

TABLE 4.1   Selective Wetting of Metal Surfaces

	Unoxidized steel St 3		Oxidized steel St 3		Unoxidized copper		Oxidized coper	
	$\cos \theta$	W_a	$\cos \theta$	W_a	$\cos \theta$	W_a	$\cos \theta$	W_a
Water	−0.52	21.15	−0.69	13.8w	−0.89	4.9	−0.98	0.89
Phenol	0.59	20.54	0.5	19.35	−0.68	4.13	−0.72	3.61
Formamide	−0.66	10.38	−0.33	20.02	−0.73	8.32	−0.86	4.31
DMFA	−0.46	5.25	0.15	11.27	−0.59	4.02	−0.63	3.63
Glycerol	−0.65	12.86	−0.71	10.59	−0.77	8.4	−0.89	4.02
DMSO	−0.19	13.06	−0.21	12.72	−0.28	11.59	−0.85	2.42

by neutral hydrocarbons, with the substrates being hydrophobic. A low work of adhesion between metal (metal oxide) and the droplet of polar liquid under hydrocarbon may point to formation of a film of that hydrocarbon on the surface.

Calculation of the γ^{LW}, γ^{+} and γ^{-} for steel surfaces by the algorithm described in detail in Refs. [11, 17] is shown in Table 4.2. According to the results, it can be stated that the polarity of metal substrates is weakly pronounced, which is evidenced by a small (in all cases) value of the SFE acid-base component. In addition, it is clear from Tables 4.1 and 4.2 that the values significantly differ for the oxidized and unoxidized surfaces. For the copper surface, due to its poor wetting with test liquids, a correct determination of the SFE components and parameters was not possible. The values obtained for the total SFE do not exceed 10 mN/m, which makes no physical sense.

Basing on the above stated, the following can be concluded:

- The proposed approach is an interesting and promising line in scientific research, since the contact angle values of wetting in the air do not completely characterize the molecular nature of metal surfaces;

TABLE 4.2   SFE Component and Parameter Values of St3 Steel Plates

	γ_s	γ^{LW}	γ^{AB}	γ^{+}	γ^{-}
Unoxidized steel St 3	38.94	38.53	0.41	1.33	0.03
Oxidized steel St 3	58.38	58.29	0.09	0.003	0.66

- Using the method of selective wetting can give the more correct information, the more SFE of neutral hydrocarbon differs from that of metal;
- The Method of selective wetting can be used to monitor the formation of thin oxide films on the surface of metals as well as different adsorption processes, which is important in the study of adhesive interaction in polymer- metal joints. There is also a way to estimate formation/thickness of the ultrathin film through the calculations of disjoining pressure.

This work was funded by the subsidy of the Russian Government to support the Program of competitive growth of Kazan Federal University among world-class academic centers and universities.

KEYWORDS

- **hydrophilicity**
- **hydrophobicity**
- **metal substrate**
- **selective wetting**
- **surface free energy**
- **work of adhesion**

REFERENCES

1. Tammann, G., Tomke, R. Die Abhangigkeit der OberflSchenspannung und die Warme vor Glassen, Z. anorg. allgem. Chem. V. I 62, No 1, 1927.
2. Lazarev S. Yu. Evaluation of the properties of substances according to the surface energy and hardness criteria. Metalloobrabotka, 2003, No. 2, 38–42.
3. Obreimov I. V. Determination of the refraction index using no devices. Trudy GOI, 1923. V.3. No. 17. 19–34.
4. Quick reference book of physical-chemical magnitudes. 8th ed. Ed. A. A. Ravdel and A. M. Ponomarev, L.: Khimiya, 1983. 20–21.
5. Starostina I. A., Sokorova N. V., Stiyanov O. V., Khakimullin Yu.N., Kurbangaleeva A. R. Relationship between adhesive and acid-base interactions. Vestnik Kazanskogo Technol. Univ. 2012. No 8. 132–134.

6. Rebinder P. A. Physicochemistry of flotation processes. M.: Metallurgizdat, 1933. 230 p.
7. Summ B. D. Basics of colloid chemistry. Publishing Centre "Akademiya," 2007.
8. Voyutsky C. C. Program for colloid chemistry. Khimiya, 1975, 512 p.
9. Abdullin I. G. Corrosion of oil-and-gas and oil-field equipment, 1990, 72 p.
10. Plugatyr V. I. Corrosion of metal constructions and protective coatings in hydrogen sulfide media. 2004, 128 p.
11. Nguyen D. A., Starostina I. A., Stoyanov O. V., Ivanova A. A. Determination of thermodynamic characteristics of metal surfaces upon selective wetting// Vestnik Kazanskogo Technol. Univ. 2013. No 18, 164–167.
12. Starostina I. A., Nguyen D. A., Burdova E. V., Stoyanov O. V. Adhesion of polymers to metals: new approaches to the estimation of the surface properties of metals. Klei. Germetiki. Tekhnologii. 2012. No.7. 27–30.
13. McCaffertyE. Acid- base effects in polymer adhesion at metal surfaces/McCaffertyE. J. AdhesionSci. Technol. 2002. V. 16. №3. 239–255.
14. Starostina A., Nguyen D. A., Burdova E. V., Stoyanov O. V. Estimation of acid-base properties of metal substrate surfaces. Vestnik Kazanskogo Technol. Univ. 2012. No 5, P.57–61.
15. Della Volpe C., Siboni S. Acid- base surface free energies of solids and the definition of scales in the Good – van Oss – Chaudhury theory. J. Adhesion Sci. Technol. 2000. V. 14. №.2. 235–272.
16. Klinov I. Ya. Corrosion prevention in the chemical and oil refining industries.1967, 208 p.
17. Starostina I. A., Stoyanov O. V., Sokorova N. V. Determination of surface free energy parameters by the spatial method. Klei. Germetiki. Tekhnologii. 2012. No 11, 31–33.

CHAPTER 5

TECHNICAL NOTES IN APPLIED QUANTUM CHEMISTRY

V. A. BABKIN,[1] D. S. ANDREEV,[1] YU. A. PROCHUKHAN,[2]
K. YU. PROCHUKHAN,[2] and G. E. ZAIKOV[3]

[1]*Volgograd State Architect-build University, Sebrykov Department, Russia*

[2]*Bashkir State University, Kommunisticheskaya ul., 19, Ufa, Respublika Bashkortostan, 450076, Russia*

[3]*Institute of Biochemical Physics, Russian Academy of Sciences, Russia*

CONTENTS

5.1 THEORETICAL ESTIMATION OF ACID FORCE OF MOLECULE P-DIMETHOXY-TRANS-STILBENE BY METHOD AB INITIO

5.1.1 INTRODUCTION

In this section, for the first time quantum chemical calculation of a molecule of p-dimethoxy-trans-stilbene is executed by method AB initio with optimization of geometry on all parameters. The optimized geometrical and electronic structure of this compound is received. Acid power of p-dimethoxy-trans-stilbene is theoretically appreciated. It is established, than it to relate to a class of very weak H-acids (pKa=+36, where pKa-universal index of acidity).

The aim of this work is a study of electronic structure of molecule p-dimethoxy-trans-stilbene [1] and theoretical estimation its acid power by quantum-chemical method AB initio in base 6–311G**. The calculation was done with optimization of all parameters by standard gradient method built-in in PC GAMESS [2]. The calculation was executed in approach the insulated molecule in gas phase. Program MacMolPlt was used for visual presentation of the model of the molecule. [3].

5.1.2 METHODICAL PART

Geometric and electronic structures, general and electronic energies of molecule p-dimethoxy-trans-stilbene was received by method AB INITIO in base 6–311G** and are shown on Fig. 5.1 and in Table 5.1. The universal factor of acidity was calculated by formula: pKa = $49.04 - 134.6 * q_{max}^{H+}$ [4, 5] (where, q_{max}^{H+} – a maximum positive charge on atom of the hydrogen $q_{max}^{H+} = +0.10$ (for p-dimethoxy-trans-stilbene q_{max}^{H+} alike Table 5.1)). This same formula is used in references [6]. pKa=36.

Quantum-chemical calculation of molecule p-dimethoxy-trans-stilbene by method AB INITIO in base 6–311G** was executed for the first time. Optimized geometric and electronic structure of thise compound was received. Acid power of molecule p-dimethoxy-trans-stilbene was theoretically evaluated (pKa=36). Thise compound pertain to class of very weak H-acids (pKa>14).

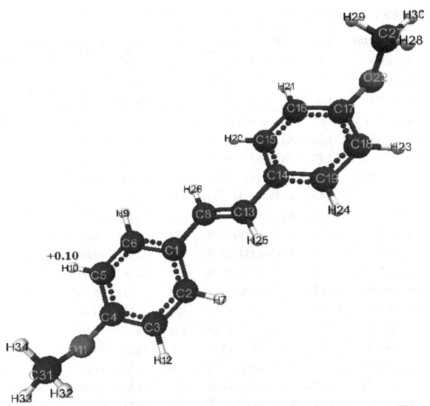

FIGURE 5.1 Geometric and electronic molecule structure of *p*-dimethoxy-trans-stilbene ($E_0 = -2,005,131$ kDg/mol, $E_{el} = -4,981,689$ kDg/mol).

TABLE 5.1 Optimized Bond Lengths, Valence Corners and Charges on Atoms of the Molecule *p*-Dimethoxy-Trans-Stilbene

Bond lengths	R, A	Valence corners	Grad	Atom	Charges on atoms
C(2)-C(1)	1.40	C(5)-C(6)-C(1)	122	C(1)	−0.05
C(3)-C(2)	1.38	C(1)-C(2)-C(3)	121	C(2)	−0.07
C(4)-C(3)	1.39	C(2)-C(3)-C(4)	120	C(3)	−0.09
C(5)-C(4)	1.38	C(3)-C(4)-C(5)	120	C(4)	+0.21
C(6)-C(5)	1.38	O(11)-C(4)-C(5)	120	C(5)	−0.09
C(6)-C(1)	1.39	C(4)-C(5)-C(6)	120	C(6)	−0.07
H(7)-C(2)	1.07	C(2)-C(1)-C(6)	118	H(7)	+0.09
C(8)-C(1)	1.48	C(1)-C(2)-H(7)	120	C(8)	−0.08

TABLE 5.1 (Continued)

Bond lengths	R, A	Valence corners	Grad	Atom	Charges on atoms
H(9)-C(6)	1.08	C(2)-C(1)-C(8)	123	H(9)	+0.09
H(10)-C(5)	1.07	C(14)-C(13)-C(8)	127	H(10)	+0.10
O(11)-C(4)	1.36	C(5)-C(6)-H(9)	119	O(11)	−0.49
H(12)-C(3)	1.08	C(4)-C(5)-H(10)	119	H(12)	+0.10
C(13)-C(8)	1.33	C(3)-C(4)-O(11)	120	C(13)	−0.08
C(13)-C(14)	1.48	C(2)-C(3)-H(12)	121	C(14)	−0.06
C(14)-C(19)	1.39	C(1)-C(8)-C(13)	126	C(15)	−0.06
C(15)-C(14)	1.40	C(19)-C(14)-C(13)	119	C(16)	−0.09
C(16)-C(15)	1.38	C(15)-C(14)-C(13)	124	C(17)	+0.21
C(17)-C(16)	1.39	C(18)-C(19)-C(14)	122	C(18)	−0.09
C(18)-C(17)	1.38	C(19)-C(14)-C(15)	118	C(19)	−0.07
C(19)-C(18)	1.38	C(14)-C(15)-C(16)	121	H(20)	+0.09
H(20)-C(15)	1.07	C(15)-C(16)-C(17)	120	H(21)	+0.10
H(21)-C(16)	1.08	C(16)-C(17)-C(18)	120	O(22)	−0.49
O(22)-C(17)	1.36	O(22)-C(17)-C(18)	120	H(23)	+0.10
H(23)-C(18)	1.08	C(17)-C(18)-C(19)	120	H(24)	+0.09
H(24)-C(19)	1.08	C(14)-C(15)-H(20)	120	H(25)	+0.09
H(25)-C(13)	1.08	C(15)-C(16)-H(21)	121	H(26)	+0.09
H(26)-C(8)	1.08	C(16)-C(17)-O(22)	120	C(27)	0.00
C(27)-O(22)	1.41	C(17)-C(18)-H(23)	119	H(28)	+0.08
H(28)-C(27)	1.09	C(18)-C(19)-H(24)	119	H(29)	+0.08
H(29)-C(27)	1.09	C(8)-C(13)-H(25)	119	H(30)	+0.10
H(30)-C(27)	1.08	C(1)-C(8)-H(26)	114	C(31)	0.00
C(31)-O(11)	1.41	C(17)-O(22)-C(27)	116	H(32)	+0.08
H(32)-C(31)	1.09	O(22)-C(27)-H(28)	111	H(33)	+0.10
H(33)-C(31)	1.08	O(22)-C(27)-H(29)	111	H(34)	+0.08
H(34)-C(31)	1.09	O(22)-C(27)-H(30)	107		
		C(4)-O(11)-C(31)	116		
		O(11)-C(31)-H(32)	111		
		O(11)-C(31)-H(33)	107		
		O(11)-C(31)-H(34)			

5.2   THEORETICAL ESTIMATION OF ACID FORCE OF MOLECULE P-NITRO-TRANS-STILBENE BY METHOD AB INITIO

5.2.1   INTRODUCTION

In this section, for the first time quantum chemical calculation of a molecule of p-nitro-trans-stilbene is executed by method AB INITIO with optimization of geometry on all parameters. The optimized geometrical and electronic structure of this compound is received. Acid power of p-nitro-trans-stilbene is theoretically appreciated. It is established, than it to relate to a class of very weak H-acids (pKa=+29, where pKa-universal index of acidity).

The aim of this work is a study of electronic structure of molecule p-nitro-trans-stilbene [1] and theoretical estimation its acid power by quantum-chemical method AB INITIO in base 6–311G**. The calculation was done with optimization of all parameters by standard gradient method built-in in PC GAMESS [2]. The calculation was executed in approach the insulated molecule in gas phase. Program MacMolPlt was used for visual presentation of the model of the molecule. [3].

5.2.2   METHODICAL PART

Geometric and electronic structures, general and electronic energies of molecule p-nitro-trans-stilbene was received by method AB initio in base 6–311G** and are shown on Figure 5.2 and in Table 5.2. The universal factor of acidity was calculated by formula: $pKa = 49.04 - 134.6 * q_{max}^{H+}$ [4,5] (where, q_{max}^{H+} – a maximum positive charge on atom of the hydrogen $q_{max}^{H+}=+0.15$ (for p-nitro-trans-stilbene q_{max}^{H+} alike Table 5.2)). This same formula is used in references [6]. pKa=29.

Quantum-chemical calculation of molecule p-nitro-trans-stilbene by method AB INITIO in base 6–311G** was executed for the first time. Optimized geometric and electronic structure of thise compound was received. Acid power of molecule p-nitro-trans-stilbene was theoretically evaluated (pKa=29). Thise compound pertain to class of very weak H-acids (pKa>14).

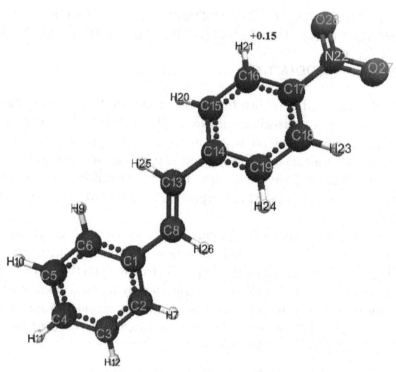

FIGURE 5.2   Geometric and electronic molecule structure of p-nitro-trans-stilbene (E_0 = −1,941,473 kDg/mol, E_{el} = −4,613,413 kDg/mol).

TABLE 5.2   Optimized Bond Lengths, Valence Corners and Charges on Atoms of the Molecule p-Nitro-Trans-Stilbene

Bond lengths	R, A	Valence corners	Grad	Atom	Charges on atoms
C(2)-C(1)	1.39	C(5)-C(6)-C(1)	121	C(1)	−0.06
C(3)-C(2)	1.38	C(1)-C(2)-C(3)	121	C(2)	−0.08
C(4)-C(3)	1.38	C(2)-C(3)-C(4)	120	C(3)	−0.09
C(5)-C(4)	1.39	C(3)-C(4)-C(5)	120	C(4)	−0.09
C(6)-C(5)	1.38	C(4)-C(5)-C(6)	120	C(5)	−0.09
C(6)-C(1)	1.39	C(2)-C(1)-C(6)	118	C(6)	−0.07
H(7)-C(2)	1.08	C(1)-C(2)-H(7)	119	H(7)	+0.09
C(8)-C(1)	1.48	C(2)-C(1)-C(8)	119	C(8)	−0.05
H(9)-C(6)	1.07	C(14)-C(13)-C(8)	126	H(9)	+0.09
H(10)-C(5)	1.08	C(5)-C(6)-H(9)	119	H(10)	+0.10

TABLE 5.2 (Continued)

Bond lengths	R, A	Valence corners	Grad	Atom	Charges on atoms
H(11)-C(4)	1.08	C(1)-C(6)-H(9)	120	H(11)	+0.10
H(12)-C(3)	1.08	C(4)-C(5)-H(10)	120	H(12)	+0.10
C(13)-C(8)	1.33	C(3)-C(4)-H(11)	120	C(13)	−0.09
C(13)-C(14)	1.48	C(2)-C(3)-H(12)	120	C(14)	−0.02
C(14)-C(19)	1.40	C(1)-C(8)-C(13)	127	C(15)	−0.10
C(15)-C(14)	1.39	C(19)-C(14)-C(13)	123	C(16)	−0.02
C(16)-C(15)	1.38	C(15)-C(14)-C(13)	119	C(17)	+0.06
C(17)-C(16)	1.38	C(18)-C(19)-C(14)	121	C(18)	−0.02
C(18)-C(17)	1.38	C(19)-C(14)-C(15)	118	C(19)	−0.09
C(19)-C(18)	1.38	C(14)-C(15)-C(16)	121	H(20)	+0.10
H(20)-C(15)	1.07	C(15)-C(16)-C(17)	119	H(21)	+0.15
H(21)-C(16)	1.07	C(16)-C(17)-C(18)	122	N(22)	+0.39
N(22)-C(17)	1.46	N(22)-C(17)-C(18)	119	H(23)	+0.15
H(23)-C(18)	1.07	C(17)-C(18)-C(19)	119	H(24)	+0.10
H(24)-C(19)	1.07	C(14)-C(15)-H(20)	120	H(25)	+0.10
H(25)-C(13)	1.08	C(15)-C(16)-H(21)	121	H(26)	+0.10
H(26)-C(8)	1.08	C(16)-C(17)-N(22)	119	O(27)	−0.38
O(27)-N(22)	1.19	C(17)-C(18)-H(23)	120	O(28)	−0.38
O(28)-N(22)	1.19	C(18)-C(19)-H(24)	119		
		C(8)-C(13)-H(25)	120		
		C(1)-C(8)-H(26)	114		
		C(17)-N(22)-O(27)	118		
		C(17)-N(22)-O(28)	118		

5.3 THEORETICAL ESTIMATION OF ACID FORCE OF MOLECULE α-CYCLOPROPYL-*P*-ISOPROPYLSTYRENE BY METHOD AB INITIO

5.3.1 INTRODUCTION

In this section, for the first time quantum chemical calculation of a molecule of α-cyclopropyl-*p*-isopropylstyrene is executed by method AB initio with optimization of geometry on all parameters. The optimized geometrical and electronic structure of this compound is received. Acid power of α-cyclopropyl-p-izopropylstyrene is theoretically appreciated. It is established, than it to relate to a class of very weak H-acids (pKa=+33, where pKa-universal index of acidity).

The aim of this work is a study of electronic structure of molecule α-cyclopropyl-p-isopropylstyrene [1] and theoretical estimation its acid power by quantum-chemical method AB Initio in base 6–311G**. The calculation was done with optimization of all parameters by standard gradient method built-in in PC GAMESS [2]. The calculation was executed in approach the insulated molecule in gas phase. Program MacMolPlt was used for visual presentation of the model of the molecule. [3].

5.3.2 METHODICAL PART

Geometric and electronic structures, general and electronic energies of molecule α-cyclopropyl-p-isopropylstyrene was received by method AB INITIO in base 6–311G** and are shown on Figure 5.3 and in Table 5.3. The universal factor of acidity was calculated by formula: pKa = 49.04– 134.6*q_{max}^{H+} [4,5] (where, q_{max}^{H+} – a maximum positive charge on atom of the hydrogen q_{max}^{H+}=+0.12 (for α-cyclopropyl-p-isopropylstyrene q_{max}^{H+} alike Table 5.3)). This same formula is used in references [6]. pKa=33.

Quantum-chemical calculation of molecule α-cyclopropyl-p-isopropylstyrene by method AB initio in base 6–311G** was executed for the first time. Optimized geometric and electronic structure of thise compound was received. Acid power of molecule α-cyclopropyl-p-isopropylstyrene was theoretically evaluated (pKa=33). Thise compound pertain to class of very weak H-acids (pKa>14).

5.4 THEORETICAL ESTIMATION OF ACID FORCE OF MOLECULE α-CYCLOPROPYL-2,4-DIMETHYLSTYRENE BY METHOD AB INITIO

5.4.1 INTRODUCTION

In this section, for the first time quantum chemical calculation of a molecule of α-cyclopropyl-2,4-dimethylstyrene is executed by method AB INITIO with optimization of geometry on all parameters. The optimized

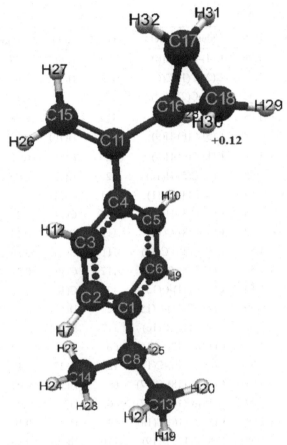

FIGURE 5.3 Geometric and electronic molecule structure of α-cyclopropyl-*p*-isopropylstyrene ($E_0 = -1,417,130$ kDg/mol, $E_{el} = -3,604,055$ kDg/mol).

TABLE 5.3 Optimized Bond Lengths, Valence Corners and Charges on Atoms of the Molecule α-Cyclopropyl-*p*-Isopropylstyrene

Bond lengths	R, A	Valence corners	Grad	Atom	Charges on atoms
C(2)-C(1)	1.39	C(5)-C(6)-C(1)	121	C(1)	−0.10
C(3)-C(2)	1.38	C(1)-C(2)-C(3)	121	C(2)	−0.06
C(4)-C(3)	1.39	C(2)-C(3)-C(4)	121	C(3)	−0.07
C(5)-C(4)	1.39	C(3)-C(4)-C(5)	118	C(4)	−0.03
C(6)-C(5)	1.38	C(11)-C(4)-C(5)	121	C(5)	−0.06

TABLE 5.3   (Continued)

Bond lengths	R, A	Valence corners	Grad	Atom	Charges on atoms
C(6)-C(1)	1.39	C(4)-C(5)-C(6)	121	C(6)	−0.08
H(7)-C(2)	1.08	C(2)-C(1)-C(6)	117	H(7)	+0.09
C(8)-C(1)	1.52	C(1)-C(2)-H(7)	120	C(8)	−0.15
H(9)-C(6)	1.08	C(2)-C(1)-C(8)	122	H(9)	+0.08
H(10)-C(5)	1.08	C(5)-C(6)-H(9)	119	H(10)	+0.08
C(11)-C(4)	1.50	C(1)-C(6)-H(9)	120	C(11)	−0.11
H(12)-C(3)	1.08	C(4)-C(5)-H(10)	120	H(12)	+0.09
C(13)-C(8)	1.54	C(3)-C(4)-C(11)	121	C(13)	−0.20
C(14)-C(8)	1.53	C(2)-C(3)-H(12)	119	C(14)	−0.21
C(15)-C(11)	1.32	C(1)-C(8)-C(13)	112	C(15)	−0.14
C(16)-C(11)	1.50	C(14)-C(8)-C(13)	111	C(16)	−0.16
C(17)-C(16)	1.50	C(1)-C(8)-C(14)	112	C(17)	−0.21
C(17)-C(18)	1.50	C(4)-C(11)-C(15)	121	C(18)	−0.17
C(18)-C(16)	1.50	C(16)-C(11)-C(15)	122	H(19)	+0.09
H(19)-C(13)	1.09	C(4)-C(11)-C(16)	117	H(20)	+0.10
H(20)-C(13)	1.09	C(18)-C(17)-C(16)	60	H(21)	+0.08
H(21)-C(13)	1.09	C(11)-C(16)-C(17)	122	H(22)	+0.10
H(22)-C(14)	1.09	C(16)-C(18)-C(17)	60	H(23)	+0.09
H(23)-C(14)	1.09	C(18)-C(16)-C(17)	60	H(24)	+0.08
H(24)-C(14)	1.09	C(11)-C(16)-C(18)	120	H(25)	+0.10
H(25)-C(8)	1.09	C(8)-C(13)-H(19)	111	H(26)	+0.10
H(26)-C(15)	1.08	C(8)-C(13)-H(20)	111	H(27)	+0.10
H(27)-C(15)	1.07	C(8)-C(13)-H(21)	111	H(28)	+0.12
H(28)-C(16)	1.08	C(8)-C(14)-H(22)	111	H(29)	+0.11
H(29)-C(18)	1.08	C(8)-C(14)-H(23)	110	H(30)	+0.12
H(30)-C(18)	1.08	C(8)-C(14)-H(24)	112	H(31)	+0.11
H(31)-C(17)	1.08	C(1)-C(8)-H(25)	107	H(32)	+0.11
H(32)-C(17)	1.07	C(11)-C(15)-H(26)	122		
		C(11)-C(15)-H(27)	122		
		C(11)-C(16)-H(28)	114		
		C(16)-C(18)-H(29)	118		
		C(16)-C(18)-H(30)	117		

TABLE 5.3 (Continued)

Bond lengths	R, A	Valence corners	Grad	Atom	Charges on atoms
		C(16)-C(17)-H(31)	118		
		C(18)-C(17)-H(31)	118		
		C(16)-C(17)-H(32)	119		

geometrical and electronic structure of this compound is received. Acid power of α-cyclopropyl-2,4-dimethylstyrene is theoretically appreciated. It is established, than it to relate to a class of very weak H-acids (pKa=+33, where pKa-universal index of acidity).

The aim of this work is a study of electronic structure of molecule α−cyclopropyl-2,4-dimethylstyrene [1] and theoretical estimation its acid power by quantum-chemical method AB INITIO in base 6–311G**. The calculation was done with optimization of all parameters by standard gradient method built-in in PC GAMESS [2]. The calculation was executed in approach the insulated molecule in gas phase. Program MacMolPlt was used for visual presentation of the model of the molecule [3].

5.4.2 METHODICAL PART

Geometric and electronic structures, general and electronic energies of molecule α−cyclopropyl-2,4-dimethylstyrene was received by method AB INITIO in base 6–311G** and are shown on Figure 5.4 and in Table 5.4. The universal factor of acidity was calculated by formula: pKa = $49.04 - 134.6*q_{max}^{H+}$ [4,5] (where, q_{max}^{H+} – a maximum positive charge on atom of the hydrogen $q_{max}^{H+} = +0.12$ (for α−cyclopropyl-2,4-dimethylstyrene q_{max}^{H+} alike Table 5.4)). This same formula is used in references [6] pKa=33.

Quantum-chemical calculation of molecule α−cyclopropyl-2,4-dimethylstyrene by method AB INITIO in base 6–311G** was executed for the first time. Optimized geometric and electronic structure of this compound was received. Acid power of molecule α−cyclopropyl-2,4-dimethylstyrene was theoretically evaluated (pKa=33). Thise compound pertain to class of very weak H-acids (pKa>14).

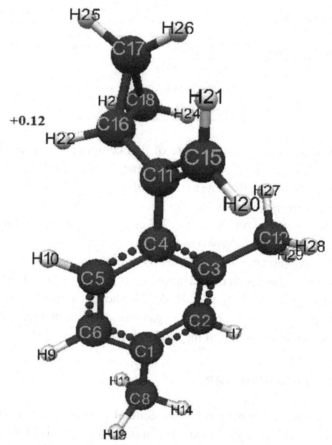

FIGURE 5.4 Geometric and electronic molecule structure of α-cyclopropyl-2,4-dimethylstyrene ($E_0 = -1,316,876$ kDg/mol, $E_{el} = -3,287,978$ kDg/mol)

TABLE 5.4 Optimized Bond Lengths, Valence Corners and Charges on Atoms of the Molecule α-Cyclopropyl-2,4-Isopropylstyrene

Bond lengths	R, A	Valence corners	Grad	Atom	Charges on atoms
C(2)-C(1)	1.38	C(5)-C(6)-C(1)	120	C(1)	−0.11
C(3)-C(2)	1.39	C(1)-C(2)-C(3)	123	C(2)	−0.07
C(4)-C(3)	1.40	C(2)-C(3)-C(4)	119	C(3)	−0.11
C(5)-C(4)	1.39	C(12)-C(3)-C(4)	123	C(4)	−0.04
C(6)-C(5)	1.38	C(3)-C(4)-C(5)	119	C(5)	−0.04

TABLE 5.4   (Continued)

Bond lengths	R, A	Valence corners	Grad	Atom	Charges on atoms
C(6)-C(1)	1.39	C(11)-C(4)-C(5)	118	C(6)	−0.09
H(7)-C(2)	1.08	C(4)-C(5)-C(6)	122	H(7)	+0.07
C(8)-C(1)	1.51	C(2)-C(1)-C(6)	118	C(8)	−0.18
H(9)-C(6)	1.08	C(1)-C(2)-H(7)	119	H(9)	+0.08
H(10)-C(5)	1.08	C(2)-C(1)-C(8)	121	H(10)	+0.08
C(11)-C(4)	1.50	C(5)-C(6)-H(9)	120	C(11)	−0.16
C(12)-C(3)	1.51	C(1)-C(6)-H(9)	120	C(12)	−0.16
H(13)-C(8)	1.09	C(4)-C(5)-H(10)	119	H(13)	+0.11
H(14)-C(8)	1.08	C(3)-C(4)-C(11)	123	H(14)	+0.09
C(15)-C(11)	1.32	C(18)-C(16)-C(11)	121	C(15)	−0.13
C(16)-C(11)	1.50	C(2)-C(3)-C(12)	119	C(16)	−0.14
C(16)-C(18)	1.50	C(1)-C(8)-H(13)	111	C(17)	−0.21
C(17)-C(16)	1.50	C(1)-C(8)-H(14)	111	C(18)	−0.17
C(18)-C(17)	1.50	C(4)-C(11)-C(15)	122	H(19)	+0.10
H(19)-C(8)	1.09	C(16)-C(11)-C(15)	123	H(20)	+0.10
H(20)-C(15)	1.08	C(4)-C(11)-C(16)	115	H(21)	+0.10
H(21)-C(15)	1.08	C(17)-C(18)-C(16)	60	H(22)	+0.12
H(22)-C(16)	1.08	C(11)-C(16)-C(17)	123	H(23)	+0.11
H(23)-C(18)	1.08	C(18)-C(16)-C(17)	60	H(24)	+0.12
H(24)-C(18)	1.08	C(16)-C(17)-C(18)	60	H(25)	+0.11
H(25)-C(17)	1.08	C(1)-C(8)-H(19)	111	H(26)	+0.11
H(26)-C(17)	1.07	C(11)-C(15)-H(20)	121	H(27)	+0.10
H(27)-C(12)	1.08	C(11)-C(15)-H(21)	122	H(28)	+0.11
H(28)-C(12)	1.09	C(11)-C(16)-H(22)	113	H(29)	+0.09
H(29)-C(12)	1.08	C(18)-C(16)-H(22)	114		
		C(17)-C(18)-H(23)	119		
		C(17)-C(18)-H(24)	117		
		C(16)-C(17)-H(25)	118		
		C(16)-C(17)-H(26)	119		
		C(3)-C(12)-H(27)	112		
		C(3)-C(12)-H(28)	111		
		C(3)-C(12)-H(29)	110		

KEYWORDS

- acid power
- method AB Initio
- *p*-dimethoxy-trans-stilbene
- quantum chemical calculation

REFERENCES

1. Kennedi J. Cationic polimerization of olefins. Moscow, 1978. 431 p.
2. M. W. Shmidt, K. K. Baldrosge, J. A. Elbert, M. S. Gordon, J. H. Enseh, S. Koseki, N. Matsvnaga., K. A. Nguyen, S. J. Su, et al. J. Comput. Chem.14, 1347–1363 (1993).
3. B. M. Bode, M. S. Gordon J. Mol. Graphics Mod., 16, 1998, 133–138.
4. V. A. Babkin, R. G. Fedunov, K. S. Minsker et al. Oxidation communication, 2002, №1, 25, 21–47.
5. V. A. Babkin et al. Oxidation communication, 21, №4, 1998, 454–460.
6. V. A. Babkin, G. E. Zaikov. Nobel laureates and nanotechnology of the applied quantum chemistry. USA. New York. Nova Science Publisher. 2010. pp. 351.

CHAPTER 6

NEW INSIGHTS IN NANOPOROUS MEMBRANE SCIENCE

B. HADAVI MOGHADAM and A. K. HAGHI

Department of Textile Engineering, University of Guilan, Rasht, Iran; E-mail: AKHaghi@yahoo.com

CONTENTS

ABSTRACT

Nanoporous membranes are an important class of nanomaterials that can be used in many applications, especially in micro and nanofiltration. Electrospun nanofibrous membranes have gained increasing attention due to the high porosity, large surface area per mass ratio along with small pore sizes, flexibility, and fine fiber diameter, and their production and application in development of filter media. Image analysis is a direct and accurate technique that can be used for characterization of porous media. This technique, due to its convenience in detecting individual pores in a porous media, has some advantages for pore measurement. The three-dimensional reconstruction of porous media, from the information obtained from a two-dimensional analysis of photomicrographs, is a relatively new research area. This chapter provides a detailed review on relevant approach of 3D reconstruction from two views of single 2D image. The review concisely demonstrated that 3D reconstruction consists of three steps, which is equivalent to the estimation of a specific geometry group. These steps include: estimation of the epipolar geometry existing between the stereo image pair, estimation of the affine geometry, and also camera calibration. The advantage of this system is that the 2D images do not need to be calibrated in order to obtain a reconstruction. Results for both the camera calibration and reconstruction are presented to verify that it is possible to obtain a 3D model directly from features in the images.

6.1 INTRODUCTION

Nanofibrous media have low basis weight, high permeability and small pore size that make them appropriate for a wide range of filtration applications. In addition, nanofiber membrane offers unique properties like high specific surface area (ranging from 1 to 35 m^2/g depending on the diameter of fibers), good interconnectivity of pores and potential to incorporate active chemistry or functionality on a nanoscale. Therefore, nanofibrous membranes are extensively being studied for air and liquid filtration. The performance of filtration media in all of these industries is determined by the pore structure characteristics of the media. There are several well documented approaches in literature to evaluate these pore structure, properties including scanning electron microscopy (SEM) analysis, gas pycnometry and adsorption, flow and mercury porosimetry, theoretical modeling, and most recently micro computed tomography (Micro-CT). Because of the advantages and disadvantages of Correspondence current approaches, virtues and pitfalls of each technique should be scrutinized. Sometimes a combination of techniques is required. However, a single nondestructive and capable of providing a comprehensive set of data is the most attractive option. These techniques are also used for pore structure characterization of nanofibrous membranes, but the low stiffness and high pressure sensitivity of nanofiber mats limit the application of these techniques, because these methods cannot be used for porosity measurement of various surface layers and can only measure the total porosity of nanofiber mat. Therefore, an accurate estimation of porosity in these grades of materials (Nanofiber mat) is a difficult task [1–6].

Porous media typically contain an interconnected three-dimensional (3D) network of channels of nonuniform size and shape (Fig. 6.1). Usually porosity is determined for materials with a three-dimensional structure, for example, relatively thick nonwoven fabrics. Nevertheless, for two-dimensional (2D) textiles such as woven fabrics and relatively thin nonwovens it is often assumed that porosity and POA are equal [7, 8].

Three-dimensional analysis is possible with image analysis techniques. Image-based techniques can be prohibitively tedious because enough pores must be analyzed to give an adequate statistical representation [9].

SEM analysis of electrospun fibrous membrane by incorporating different image analysis methods is one of the renowned methods for

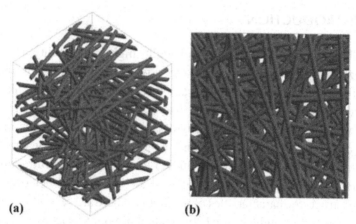

(a)   (b)

FIGURE 6.1   SEM image of electrospun nanofibrous membrane: (a) 3D (b) 2D.

researchers to measure porosity parameters of woven fabric, nonwoven and membranes, and nanofiber membrane. Although image analysis of SEM micrographs for geometrical characterization is useful for measuring the total porosity, pore shape, pore size and pore size distribution of relatively thin nonwovens, it cannot be applied to multilayer electrospun fibrous analysis. Another problem encounter to this method is that it is not possible to measure 3D pore characteristics of the membrane and it is limited on relatively small fields of view [4].

Three-dimensional analysis is possible with image analysis techniques. Image-based techniques can be prohibitively tedious because enough pores must be analyzed to give an adequate statistical representation. 3D pore structures of nanofibrous membrane have been evaluated by nondestructive three-dimensional laser scanning confocal microscope (LSCM) and micro computed tomography (Micro-CT), 3D electron backscatter diffraction (EBSD), nuclear magnetic resonance imaging [4–19].

The topic of obtaining 3D models from images is a fairly new research field of computer vision.

A set of techniques for creating a 3D representation of a view from one or more 2D images can be used to Image-based modeling. A solution to this lack of 3D content is to convert existing 2D material to 3D [20].

3D reconstruction is a challenging task in computer vision and has received considerable attention recently due to the loss of a dimension (the depth of the image) in the process of photographing image and the usefulness of the recovered 3D model for a variety of applications, such as city planning, cartography, architectural design, simulations. The key task in 3D reconstruction is to recover high-quality and detailed 3D models from two or more views of the image, which may be taken from widely separated viewpoints. Due to the complexity of the images, conventional modeling techniques are very time-consuming and recreating detailed geometry become very difficult. In order to overcome these difficulties, some works have been inclined towards image-based modeling techniques, using images to drive the 3D reconstruction. However, in many image-based modeling techniques, the image is reconstructed using camera calibrated images or, when this is not the case, it is nontrivial to establish correspondences between different views of image. Image-based modeling relies on a set of techniques for creating 3D representation of a 2D image from one or more views of image [21–27].

Generally, the 3D reconstruction consists of three steps; (i) estimation of the epipolar geometry existing between the two views of the image pair, which involves feature matching in both images, (ii) estimation of the affine geometry which considered as a process of finding a special plane in projective space by means of vanishing points, and (iii) camera calibration by which it is possible to obtain a 3D model of the 2D image (Fig. 6.2) [28].

3D reconstruction of a number of perspective images is one of the fundamental problems of computer vision, while reconstruction from two views is the simplest one. To the best of our knowledge, earliest work in this field concentrated on calibrated cameras, from which it is

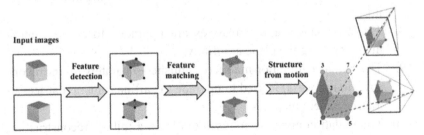

FIGURE 6.2   3D camera poses and positions.

possible to obtain a Euclidean (sometimes called metric) reconstruction of the image [29].

Finding the 2D point matches between images, which is known as the correspondence, is the first problem encountered in the process. There are many automated techniques for finding correspondences between two images, but most of them work on the basis that the same 3D point in the world (e.g., the window corner) will have a similar appearance in different images, particularly if those images are taken close together. The second challenge is how to determine where each photo was taken and the 3D location of each image point, given just a set of corresponding 2D points among the photos. If one knew the camera poses, but not the point positions, one could find the points through triangulation; conversely, if one knew the 3D points, one could find the camera poses through a process similar to triangulation called resectioning. Unfortunately, one knows, neither the camera poses nor the points [30–33].

For a 2D image pair, the individual steps of the 3D reconstruction algorithm are as follows:

1. Features are detected in each image independently.
2. A set of initial feature matches is calculated.
3. The fundamental matrix is calculated using the set of initial matches.
4. False matches are discarded and the fundamental matrix is refined.
5. Projective camera matrices are established from the fundamental matrix.
6. Vanishing points on three different planes and in three different directions are calculated from parallel lines in the images.
7. The plane at infinity is calculated from the vanishing points in both images.
8. The projective camera matrices are upgraded to affine camera matrices using the plane at infinity.
9. The camera calibration matrix (established separately for the reconstruction process) is used to upgrade the affine camera matrices to metric camera matrices.
10. Triangulation methods are used to obtain a full 3D reconstruction with the help of the metric camera matrices.

11. If needed, dense matching techniques are employed to obtain a 3D texture map of the model to be reconstructed.

Figure 6.3 shows the algorithm for individual steps of the 3D reconstruction.

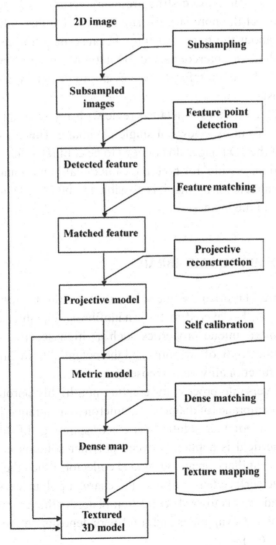

FIGURE 6.3 3D modeling framework.

This chapter will discuss a process for 3D modeling of nanofibers, as well as the benefits, limitations, construction, and performance of filters using 3D reconstruction of nanofibrous membrane. In particular, nanofibers provide marked increases in filtration efficiency at relatively small (and in some cases immeasurable) decreases in permeability. Filtration performances of the nanofibrous membranes can be defined by parameters such as pore structure characteristics of the membrane. Determination of the pore size distribution in 2D does not give general information about the pore structures because nanofibrous membranes typically include an interconnected 3D network of channels of nonuniform size and shape, therefor it is necessary to detect suitable technique for pore measurement.

This chapter provides a detailed review on relevant approach of 3D reconstruction from two views of single 2D image. The advantage of this system is that the 2D images do not need to be calibrated in order to obtain a reconstruction. Results for both the camera calibration and reconstruction are presented to verify that it is possible to obtain a 3D model directly from features in the images.

6.2   NANOFIBROUS MEMBRANE

Nanofibers are classified as fibers less than 1 micrometer in diameter. These fibers can be layered to form nanofibrous membranes, and these membranes offer unique properties such as high specific surface area, good interconnectivity of the pores and the potential to incorporate active chemistry or functionality on a nanoscale.

The most versatile process for creating non-highly porous nonwoven nanofibrous membrane (of diameter submicron to nanorange 100–500 nm) from variety of polymer solutions is electrospinning. In this process, a strong electric field is applied between polymer solution contained in a syringe with a capillary tip and grounded collector. When the electric field overcomes the surface tension force, the charged polymer solution forms a liquid jet and travels towards collection plate. As the jet travels through the air, the solvent evaporates and dry fibers deposits on the surface of a collector (Fig. 6.4).

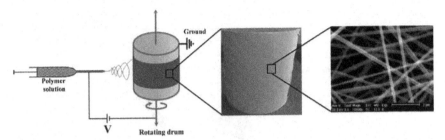

FIGURE 6.4   Schematic representation of electrospinning process.

The morphology and the structure of the electrospun nanofibrous membrane are dependent upon many parameters which are mainly divided into three categories: solution properties (the concentration, liquid viscosity, surface tension, and dielectric properties of the polymer solution), processing parameters (applied voltage, volume flow rate, tip to collector distance, and the strength of the applied electric field), and ambient conditions (temperature, atmospheric pressure and humidity). Studied showed that the formation of pores on nanofibers membrane during electrospinning process affected by many parameters such as humidity, type of polymer, solvent vapor pressure, electrospinning conditions, etc. Although no generally agreed set of definitions exists, porous materials can be classified in terms of their pore sizes into various categories including capillaries (>200 nm), macropores (50–200 nm), mesopores (2–50 nm) and micropores (0.5–2 nm).

The nanofibrous membranes can be formed from a variety of polymer and polymer blends, on the screen or substrate in any size and shape to any desired thickness; size and thickness depend on the volume of solution that is electrospun and the amount of layering of the fibers. Additionally, the membrane's fiber diameter, porosity, texture, and structure can be changed by using different polymer solutions.

Electrospun nanofibrous membranes (ENMs) have high flux rates, low trans membrane pressure, low basis weight and small pore size that make them suitable for a wide range of filtration applications, also the nanofibers membrane became flexible and less hydrophobic, suitable for a wide range of bioengineering and medical applications (Fig. 6.5) [34–51].

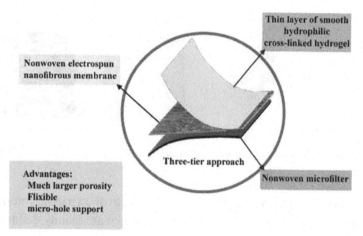

FIGURE 6.5   The nanofibrous membranes.

6.3   APPLICATION OF NANOFIBROUS MEMBRANE IN FILTRATION

Electrospun nanofibers with diameters 10–100 times smaller than those of nonwoven articles have enabled new levels of filtration performance in several diverse applications, because of their larger surface area to volume ratio and smaller pore sizes and their higher effective porosity when compared to current commercial filter media.

The membranes' smaller fiber diameters and pore sizes permit such filters to filter out more and smaller particulate. Nanofibrous membrane in the submicron range, in comparison with larger ones, is well known to provide higher filtration efficiency at the same pressure drop and relatively small decreasing in permeability in the filtration process due to decreasing filter membrane thickness and increasing applied pressure. In general controlling parameters of electrospinning allows the generation of nanofibrous membrane with different filtration characteristics.

Due to the small pore size and lack of self-supporting mechanical strength of nanofibrous filter media, the present use of its, is limited to prefiltration. Nanofibrous filter media have been constructed and tested by electrospinning a layer of synthetic polymer fibers onto melt blown or spun bonded nonwoven substrates for varied applications [52–57].

6.4   EFFECTIVE PARAMETER IN FILTRATION PROCESS

A porous media is most often characterized by its porosity and in particular by how its porosity is distributed. Other properties of the medium (e.g., permeability, tensile strength, electrical conductivity) can be connected to the porosity or to the pore structure. For this reason, many efforts have been made in an attempt to characterize the internal structure of porous media [58].

Also filtration performances of the nanofibrous membranes can be defined by parameters such as pore structure characteristics of the media such as the most constricted through pore diameter, the largest pore diameter, the mean pore diameter, pore shape, pore distribution, pore volume, pore volume distribution, surface area, liquid permeability, gas permeability and influence of operational parameters such as compressive stress, cyclic compression, pressure temperature, chemical environment, sample orientation, inhomogeneity and layered or graded structures (Fig. 6.6) [3].

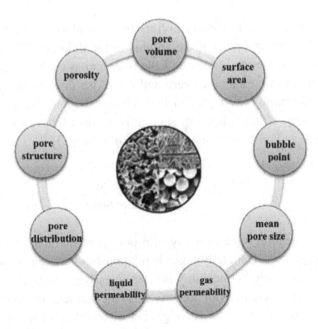

FIGURE 6.6   Pore structure characteristics.

The combination of these features implies a higher permeability (because of the large pore volume) that can withstand fouling better (because of the interconnection of the pores) and smaller pore sizes, implying that the electrospun nanostructure can support even thinner coatings that are responsible for the actual fluid filtration.

Generally three kinds of pores are found in filter media: closed pores, blind pores, and through pores. Closed pores do not allow particle to pass through the filter media, while blind pores are initially open pores that become closed within the filter media and do not allow the crossing of air. Through pores are open throughout the whole filter media depth, permitting fluid flow to move completely through the filter. These through pores and their size are most important for filtration because they allow fluid flow [51].

The pore size of a filter media is the average pore dimension through which the fluid must flow. The more efficient the filter media due to the smaller the pore enable to remove smaller particles. Performances of the newly realized membranes can be evaluated by investigation of the pore size of membranes and thus the real pore size of the commercial ones can be known [59].

Another important parameter in filter design and filter performance is Porosity. Porosity is a ratio of the volume of the fluid or void space in a filter medium to the total volume of the filter and as such it has no units and has values in the range of zero and one. Also the porosity does not give any information about pore sizes, their distribution, and their degree of connectivity. If the volume of sample is denoted by V_s, and the pore volume as $V_p = V - V_s$, we can write the porosity as [59]:

$$\varepsilon = \frac{V - V_s}{V} = \frac{V_p}{V} = \frac{Porevolume}{Totalvolume} \tag{1}$$

Also Permeability is a property of a porous media that characterizes in filtering process. Generally there is relationship between porosity and permeability. Clearly Permeability is often linearly proportional to porosity and it should increase with porosity. Also the permeability was fitted to a linear relation to the mat basis weight but was weakly related to the fiber diameter [60–63].

The Kozeny-Carman model is one of the most well known models linking porosity and permeability that considers the porous media to be made up of bundles of capillary tubes. The basic equation is [64, 65]:

$$K_{KC} = \frac{cd^2\varepsilon^3}{(1-\varepsilon)^2}$$ (2)

where:
K_{KC} = Kozeny-Carman predicted permeability, mD
c = A constant
d = Median grain size diameter, microns
ε = Effective porosity

Improved structural (pore size distribution, pore interconnectivity and porosity) and transport (permeability) properties of the electrospun filtering media can be attained by optimization of the fiber crossing and pore size, which this can be obtained by coordination of the drawing and collection rates and changing the deposition rate of nanofibers in electrospinning process [6, 34].

Electrospun nanofibrous membranes are included of fibers with diameters as low as hundreds of nanometers, and have porosities lower than 90% with pore sizes as low as several micrometers [49].

6.5 METHODS FOR THE CHARACTERIZATION OF POROSITY AND PORE STRUCTURE

Porous media play an important role in several branches of science and technology such as the petroleum and chemical industries and in medicine, biochemistry, and electrical engineering. It is important to select and appraisal of the methods of characterization of porous media rapidly, reliably [7].

Several methods can be employed to evaluate the pore characteristics of porous membranes. Generally, two kinds of techniques can be used for pore structure characterization, such as the microscopic and macroscopic techniques (Fig. 6.7).

The macroscopic techniques scan large sample areas of interest in applications. These techniques are based on permeation of a gas or a

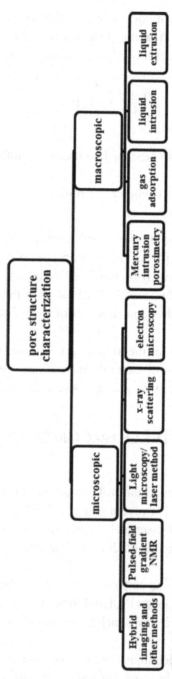

FIGURE 6.7   Various techniques for pore structure characterization of filtration media [3, 66–70].

liquid; also they can product information about pore size and distribution, but cannot give the flow properties. These techniques are inexpensive and can be fast and widely used in several applications (Table 6.1) [3,66].

Porous media are strongly characterized by their internal microstructure, which needs to be accurately described in order to determine their performance and macroscopic properties. The microscopic techniques are based on the interaction of an electro-magnetic, magnetic, or electronic radiation with the solid containing voids. Examine tiny areas, which may not be representative of macroscopic areas of interest for applications. Also none of the flow properties is determinable. The techniques are involved, time consuming and expensive. All of them characterize quantitatively the porous media internal structure.

Despite recent technological advances and the introduction of imaging techniques such as X-ray microtomography, methods to characterize

TABLE 6.1 Capabilities of the Macroscopic Techniques

Techniques		Liquid extrusion	Liquid intrusion	Gas adsorption	Mercury intrusion porosimetry
Local average porosity		√	×	×	×
Largest Constricted diameter		√	×	×	×
Many diameters of each pore		×	√	√	√
Flow distribution		√	×	×	×
volume		×	√	√	√
Volume distribution		×	√	√	√
Surface area		√	√	√	√
Operational Features	Use of toxic material	×	×	×	√
	High pressure	×	×	×	√
	Subzero temperature	×	×	√	×
	Use of fluid interest	√	√	×	×
	Involved/time consuming	×	×	√	√

quantitatively the porous media internal structure are still few and related to some specific applications. Although microtomographic image processing presents considerable difficulties, both for the intrinsic characteristics of the images and for the nature of analyzed objects [59]. Table 6.2 is summarized pore diameter range of these techniques.

Often in the past, specific analysis techniques have been applied to 2D images obtained by optical methods, but they returned only partial information about the porous media structure. Increased computational power and progresses in imaging techniques are providing researchers with the tools and data to take a big leap ahead in understanding of media microstructure, but the extension from 2D analysis 3D case represents a nontrivial challenge, requiring both a large computing effort and new methodological approaches implementation [58].

Determination of the pore size distribution in two dimensional does not give general information about the pore structures because nanofibrous membranes typically include an interconnected 3D network of channels of nonuniform size and shape, therefor it is necessary to detect suitable technique for pore measurement.

For the first time, that pycnometry is a simple and reliable method for estimating the porosity of electrospun nanofibers even for the thin coatings necessary for the regular filter media for high performance applications. The advantages of the method include: (i) minimum damage to the

TABLE 6.2   Pore Diameter and Pore Volume Measurable by Macroscopic and Microscopic Techniques

Techniques		Pore diameter range, μm
Macroscopic Techniques	Liquid extrusion	0.013–500
	Liquid intrusion	0.001–20
	Gas adsorption	0.0005–2
	Mercury intrusion porosimetry	0.03–200
Microscopic Techniques	MRI	~10 μm
	X-ray tomography (CXT)	~100 nm to 1 μm
	Electron microscopy	<~500 nm
	Light microscopy/laser method	~1 μm
	Pulsed-field gradient NMR	~10 μm or greater

fibers, (ii) chemically inert, (iii) measurement of total porosity including micro and meso-pores due to the ease of penetration of helium gas, and (iv) independent of direction of medium as the static pressure is applied. Measurement of the porosity by pycnometry allows the design of the coating structures suitable for filter media with minimum pressure drop for practical applications.

The macroscopic techniques are also used for pore structure characterization of nanofibrous membranes, but the low stiffness and high-pressure sensitivity of nanofiber mats limit application of these techniques. Image analysis was used to measure pore characteristics of woven and nonwoven geotextiles [58, 71, 72].

6.5.1   IMAGE PROCESSING

Methods used over pictures of highly polished surfaces of porous materials taken with an electron scanning microscope or optical microscope have been used to describe porous structure. Image analysis has important consequences in porous media theory for describing the fine structure of the porous space, obtaining equivalent lattice models for the real porous structure, One of the most useful results of image analysis in the study of porous structure is the three dimensional reconstruction of porous structure. The general objective of reconstructed porous media is to mimic the geometry of real media, enabling the creation of numerical realizations of the sample with the desired geometric properties [73].

Until a few years ago, the study of porous media by image analysis was limited to destructive methods such as thin sectioning, and the imaging techniques relying on those methods were inherently 2D. Indeed, the three-dimensional information of object structures was obtained by cutting the object into very thin slices, which were visualized in the light microscope, and then the two-dimensional information was interpolated into a 3D structure model. Two-dimensional reproduction from 2D multiple point statistics is not challenging because sufficient statistics are measured on a 2D training plane. Generating a 3D structure from 2D information is a truly challenging task [74, 75].

6.6   3D RECONSTRUCTION OF PORE NETWORK

A challenging question is to find a way, if it exists, to provide a realistic 3D configuration of a pore network at the right length scale. An alternative strategy to modeling the material is to perform a reconstruction of the material from limited but relatively accurate structural information about the original system, mainly 2D images of a random section and small angle scattering of the matrix. This is an inverse problem that has no general and exact solution. However, it is important to evaluate available methods able to solve such a problem with a good level of approximation. In the following, we discuss stochastic reconstructions using either correlated Gaussian fields or simulated annealing [76].

6.6.1   3D RECONSTRUCTION FROM 2D IMAGES

The problem of inferring 3D information of a scene from a set of 2D images has a long history in computer vision. This problem falls into the category of so-called inverse problems, which are prone to be ill conditioned and difficult to solve in their full generality unless additional assumptions are imposed. While, the task of selecting a correct mathematical model can be elusive, simple choices of representation can be made by exploiting geometric primitives such as points, lines, curves, surfaces, and volumes. In general, automatic 2D-to-3D media conversion algorithm recovers the missing depth information from monocular image/video input and synthesizes virtual stereoscopic view [77].

The task of converting multiple 2D images into 3D model consists of a series of processing steps:

1. Camera calibration consists of intrinsic and extrinsic parameters, without which at some level no arrangement of algorithms can work. The dotted line between Calibration and Depth determination represents that the camera calibration is usually required for determining depth.
2. Depth determination serves as the most challenging part in the whole process, as it calculates the 3D component missing from any given image – depth. The correspondence problem, finding

matches between two images so the position of the matched elements can then be triangulated in 3D space is the key issue here.

There are various depth recovery cues from single 2D image, such as linear perspective, overlapping, texture gradient and relative height. Among the pictorial depth cues, we should not fail to notice the relative height cue, which means that the closer object in real world are projected into the lower part in a 2D image plane [77].

Once you have the multiple depth maps you have to combine them to create a final mesh by calculating depth and projecting out of the camera – registration. Camera calibration will be used to identify where the many meshes created by depth maps can be combined together to develop a larger one, providing more than one view for observation. By the stage of Material Application you have a complete 3D mesh, which may be the final goal, but usually you will want to apply the color from the original photographs to the mesh. This can range from projecting the images onto the mesh randomly, through approaches of combining the textures for super resolution and finally to segmenting the mesh by material, such as specular and diffuse properties.

The most important and difficult problem in 2D-to-3D conversion is how to generate or estimate the depth information using only a single-view image. Since there is no 3D information, we should estimate relative depth differences for each region in a single-view image. Several methods have been proposed to estimate the depth information from a single-view image (Fig. 6.8).

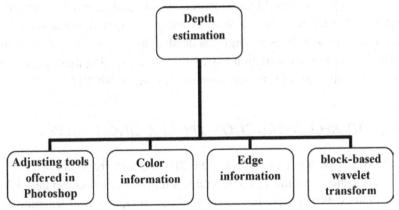

FIGURE 6.8 Various techniques for depth estimation [78].

Recovering 3D depth from images is a fundamental problem in computer vision, and has important applications in robotics, general scene understanding and 3D reconstruction. Most work on visual 3D reconstruction has focused on stereovision and on other algorithms that require multiple images, such as structure from motion and depth from defocus. These algorithms consider only the geometric (triangulation) differences. Beyond stereo/triangulation cues, there are also numerous monocular cues such as texture variations and gradients, defocus, color/haze, etc. that contain useful and important depth information.

Depth estimation from a single still image is a difficult task, since depth typically remains ambiguous given only local image features. Thus, algorithms must take into account the global structure of the image, as well as use prior knowledge about the scene. Also depth estimation is small but crucial step towards the larger goal of image understanding, in that it will help in tasks such as understanding the spatial layout of a scene, finding walkable areas in a scene, detecting objects [79].

If scene structure can be estimated then better understanding the scene can be obtained by knowing the 3D relationships between the objects within it. However, estimating structure from raw image features is notoriously difficult since local appearance is insufficient to resolve depth ambiguities. Generally, semantic understanding of a scene plays an important role in our own perception of scale and 3D structure. Producing spatially plausible 3D reconstructions of a scene from monocular images annotated with geometric cues (such as horizon, vanishing points, and surface boundaries) is a well-understood problem. However, to uniquely determine absolute depths, additional information such as texture, relative depth, and camera parameters (pose and focal length) is needed. Much recent work on automated 3D scene reconstruction has focuses on extracting these geometric cues and additional information from novel images [80, 81].

6.6.2 3D RECONSTRUCTION FROM A SINGLE VIEW

Recently the problem of Inferring the 3D structure from single view of pictured scene has attracted considerable attention in both computer vision and graphics. When there is a reference in the image, 3D information can be extracted from a single image. 3D reconstruction from a single image must necessarily be through an interactive process, where the user provides information and constraints about the scene structure. Such information may be in terms of vanishing points or lines, coplanarity, spatial interrelationship of features and camera constraints. Most methods for estimation of 3D models from a single image make strong assumptions about the scene geometry. Using geometric constraints, such as the parallelism and orthogonality of lines, and the way one building (either its exterior or interior) is structured, we can interpret the collection of the line segments [82–85].

There is a two-step approach for recovering 3D structure of outdoor images: (i) they estimate image region orientation (e.g., ground vertical) using statistical methods on image properties, such as color, texture, edge orientation, position in image, etc.; (ii) "pop-up" vertical regions by "folding" along the crease between ground and vertical regions. The Markov Random Field (MRF) represents statistical relations between scene depth and image features, learned from a training set of images with ground truth depth. Many researchers have dealt with the problem of 3D reconstruction either from a single image or from two or more images, which is more common [84].

The reconstruction problem consists of three steps, each of which is equivalent to the estimation of a specific geometry group. The first step is the estimation of the epipolar geometry that exists between the stereo image pair, a process-involving feature matching in both images. The second step estimates the affine geometry, a process of finding a special plane in projective space by means of vanishing points. Camera calibration forms part of the third step in obtaining the metric geometry, from which it is possible to obtain a 3D model of the scene.

6.7   CLASSIFICATION OF 3D GEOMETRY GROUP

Stratification of 3D vision makes it easier to perform a reconstruction. Up to this point it is possible to obtain the 3D geometry of the image, but as only a restricted number of features are extracted, it is not possible to obtain a very complete textured 3D model [28].

3D vision can be divided into four geometry groups or strata, of which Euclidean geometry is one. The simplest group is projective geometry, which forms the basis of all other groups. The other groups include affine geometry, metric geometry and then Euclidean geometry. These geometries are subgroups of each other, metric being a subgroup of affine geometry, and both these being subgroups of projective geometry [28].

Table 6.3 is summarized some characteristics of these categories.

6.7.1 PROJECTIVE GEOMETRY

The basis of most computer vision tasks, especially in the fields of 3D reconstruction from images and camera self-calibration is Algebraic and projective geometry forms [28].

In mathematics, projective geometry is an elementary nonmetrical form of geometry, that is invariant under projective transformations and it is not based on a concept of distance. This means that projective geometry has a different setting, projective space, and a selective set of basic geometric concepts in comparison to elementary geometry. In a given dimension, projective space has more points than Euclidean space and geometric transformations are allowed that move the extra points (called "points at infinity") to traditional points, and vice versa. In two dimensions, it begins

TABLE 6.3   Characteristics of 3D Geometric Transformation

Transformation	DoF	Trans. Matrix	Distortion	Invariants
Projective	15	$[\tilde{H}]_{4\times4}$		straight lines
Affine	12	$[A]_{3\times4}$		parallelism
Metric (similarity)	7	$[sR\|t]_{3\times4}$		angles
Euclidean (rigid)	6	$[R\|t]_{3\times4}$		lengths

with the study of configurations of points and lines. In higher dimensional spaces there are considered hyperplanes (that always meet), and other linear subspaces, which exhibit the principle of duality. The simplest illustration of duality is in the projective plane, where the statements "two distinct points determine a unique line" (i.e., the line through them) and "two distinct lines determine a unique point" (i.e., their point of intersection) show the same structure as propositions. Properties meaningful in projective geometry are respected by this new idea of transformation, which is more radical in its effects than expressible by a transformation matrix and translations (the affine transformations).

Theory of perspective is one source for projective geometry. Another difference from elementary geometry is the way in which parallel lines can be said to meet in a point at infinity, once the concept is translated into projective geometry's terms. Again this notion has an intuitive basis, such as railway track meeting at the horizon in a perspective drawing. See projective plane for the basics of projective geometry in two dimensions (Fig. 6.9) projective algebraic geometry (the study of projective varieties) and projective differential geometry (the study of differential invariants of the projective transformations) are two research subfields of projective geometry.

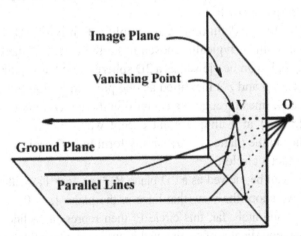

FIGURE 6.9   A two dimensional construction of perspective viewing which illustrates the formation of a vanishing point.

6.7.1.1 Projective Space and Its Homogeneous Coordinates

An n-dimensional projective space P^n is the set of one-dimensional subspaces (i.e., lines through the origin) of the vector space R^{n+1}. A point P in P^n can then be assigned homogeneous coordinates $X = [X_1, X_2, ..., X_{n+1}]^T$ among which at least one X_i is nonzero. For any nonzero $\lambda \in R$ the coordinates $Y = [\lambda X_1, \lambda X_2, ..., \lambda X_{n+1}]^T$ represent the same point P in P^n. We say that X and Y are equivalent, denoted by $X \sim Y$.

6.7.1.2 Topological Models For the Projective Space P²

In projective geometry the linear group of transformations of the line into itself contains three parameters, if we may use this convenient analytic term to stand for the geometric conception of degrees of freedom. The establishment, upon the line, of a definite number- system depends on three assumed points 0, 1, and ∞ determines completely the theory of segments and distance – the difference between two numbers giving the distance between the points to which they are affixed. The postulate of congruence, namely, that from a given point of the line and in either direction, there may be laid off upon the line one and only one segment equal to a given segment of the line, has an evident interpretation.

Figure 6.10 demonstrates two equivalent geometric interpretations of the 2D projective space P².

According to the definition, it is simply a family of 1D line {L} in R^3 through a point o (typically chosen to be the origin of the coordinate frame). Hence, P² can be viewed as a 2D sphere S² with any pair of antipodal points (e.g., p and p') identified as one point in P². On the right-hand side of Fig. 6.4, lines through the center o in the general intersection with the plane {$z = 1$} at a unique point except when they lie on the plane {$z = 0$}. Lines in the plane {$z = 0$} simply form the 1D projective space P¹ (which is in fact a circle).

Hence, P² can be viewed as a 2D plane R² (i.e., {$z = 1$}) with a circle P¹ attached. If we adopt the view that lines in the plane {$z = 0$} intersect the plane {$z = 1$} infinitely far, this circle P¹ then represents a line at infinity. Homogeneous coordinates of a point on this circle, then take the form $[x, y, 0]$; on the other hand, all regular points in R² has coordinates $[x, y, 1]^T$. In general,

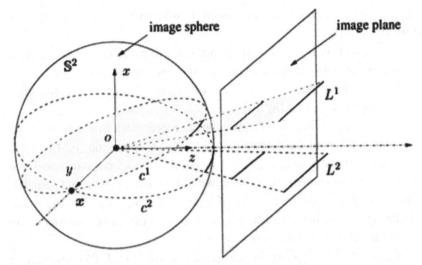

FIGURE 6.10   Topological models for projective space P^2 [86].

any projective space P^n can be visualized in a similar way: P^3 is then R^3 with a plane P^2 attached at infinity; and P^n is R^n with P^{n-1} attached at infinity.

Using this definition, R^n with its homogeneous representation can then be identified as a subset of P^n that includes exactly those points with coordinates $X = [x_1, x_2, ..., x_{n+1}]^T$ where $x_{n+1} \neq 0$. Therefore, we can always normalize the last entry to 1 by dividing X by x_{n+1} [86, 87].

6.7.2   AFFINE GEOMETRY

Affine geometry is not concerned with the notions of circle, angle and distance. It's a known dictum that in affine geometry all triangles are the same. In this context, the word affine was first used by Euler (affinis). In modern parlance, Affine Geometry is a study of properties of geometric objects that remain invariant under affine transformations (mappings).

Affine geometry is established by finding the plane at infinity in projective space for both images. The usual method of finding the plane is by determining vanishing points in both images and then projecting them into space to obtain points at infinity. Vanishing points are the intersections of two or more imaged parallel lines. This process is unfortunately very difficult to automate, as the user generally has to select the parallel lines in the images. Some automatic algorithms try to find dominant line orientations in histograms [88].

A rigid motion is an affine map, but not a linear map in general. Also, given a $m \times n$ matrix A and a vector $b \in R^m$, the set $U = \{x \in R^n \mid Ax = b\}$ of solutions of the system $Ax = b$ is an affine space, but not a vector space (linear space) in general [89].

Analytically, affine transformations are represented in the matrix form $f(x) = Ax + b$, where the determinant $\det(A)$ of a square matrix A is not 0. In a plane, the matrix is 2×2; in the space, it is 3×3. One way to arrive at the matrix representation is to select two points (two origins) and associate with each an appropriate number of independent vectors (2 in the plane, 3 in the space), to form an affine basis. b is then the translation that maps one of the selected points onto another.

A three-dimensional incidence space is a triple $(S;L;P)$ consisting of a nonempty set S (whose elements are called points) and two nonempty disjoint families of proper subsets of S, denoted by L (lines) and P (planes) respectively, which satisfy the following conditions:

1. every line (element of L) contains at least two points, and every plane (element of P) contains at least three points.
2. if x and y are distinct points of S, then there is a unique line L such that $x, y \in L$. notation. The line given by 2 is called xy.
3. if x, y and z are distinct points of S and $z \notin xy$, then there is a unique plane P such that $x, y, z \in P$.
4. if a plane P contains the distinct points x and y, then it also contains the line xy.
5. if P and Q are planes with a nonempty intersection, then $P \cap Q$ contains at least two points.

Affine transformations preserve collinearity of points: if three points belong to the same straight line, their images under affine transformations also belong to the same line and, in addition, the middle point remains between the other two points. As further examples, under affine transformations parallel lines remain parallel, concurrent lines remain concurrent (images of intersecting lines intersect), the ratio of length of line segments of a given line remains constant, the ratio of areas of two triangles remains constant (and hence the ratio of any areas remains constant), ellipses remain ellipses and the same is true for parabolas and hyperbolas (Fig. 6.11) [90].

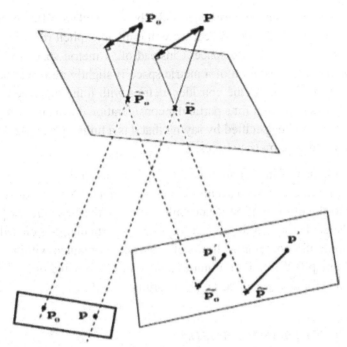

FIGURE 6.11 Affine structure under parallel projection is d_p/d_o [91].

6.7.3 METRIC GEOMETRY

This stratum corresponds to the group of similarities. The transformations in this group are Euclidean transformations such as rotation and translation. The metric stratum allows for a complete reconstruction up to an unknown scale [28].

Let X be an arbitrary set. A function $d : X \times X \to R \cup \{\infty\}$ $f_1 g$ is a metric on X if the following conditions are satisfied for all $x, y, z \in X$.

1. Positiveness: $d(x, y) \rangle 0$ if $x \neq y$, and $d(x, x) = 0$.
2. Symmetry: $d(x, y) = d(y, x)$.
3. Triangle inequality: $d(x, z) \leq d(x, y) + d(y, z)$.

A metric space is a set with a metric on it. In a formal language, a metric space is a pair (X, d) where d is a metric on X. Elements of X are called points of the metric space; $d(x, y)$ is referred to as the distance between points x and y. When the metric in question is clear from the context, we also denote the distance between x and y by $|xy|$. Unless different metrics on the same set X are considered, we will omit an explicit reference to the metric and write "a metric space X instead of "a metric space $(X; d)$." In most textbooks, the notion of a metric space is slightly narrower than our definition: traditionally one consider metrics with finite distance between points. If it is important for a particular consideration that d takes only finite values, this will be specified by saying that d is a finite metric [92, 93].

Let X and Y be metric spaces. Then,

1. a sequence in X cannot have more than one limit.
2. a point $x \in X$ is an accumulation point of a set $S \subset X$ (i.e., belongs to the closure of S) if and only if there exists a sequence $\{x_n\}_{n=1}^{\infty}$ such that $x_n \in S$ for all n and $x_n \to x$. In particular, S is closed if and only if it contains all limits of sequences contained within S.
3. a map $f: X \to Y$ is continuous at a point $x \in X$ if and only if $f(x_n) \to f(x)$ for any sequence $\{x_n\}$ converging to x [92].

6.7.4　EUCLIDEAN GEOMETRY

Geometry (from the Greek "geo" = earth and "metria" = measure) arose as the field of knowledge dealing with spatial relationships. Geometry can be split into Euclidean geometry and analytical geometry. Analytical geometry deals with space and shape using algebra and a coordinate system. Euclidean geometry deals with space and shape using a system of logical deductions. A geometry in which Euclid's fifth postulate holds, sometimes also called parabolic geometry. Two-dimensional Euclidean geometry is called plane geometry, and three-dimensional Euclidean geometry is called solid geometry. Hilbert proved the consistency of Euclidean geometry.

Euclidean geometry describes a 3D world very well (Fig. 6.12). As an example, the sides of objects have known or calculable lengths, intersecting lines determine angles between them, and lines that are parallel to a plane will never meet. But when it comes to describing the imaging process of a camera, the Euclidean geometry is not sufficient, as it is not possible to determine lengths and angles anymore, and parallel lines may intersect.

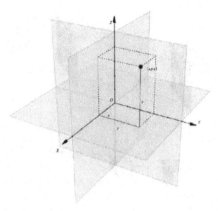

FIGURE 6.12 3D Euclidean space.

Euclidean geometry is the same as metric geometry, the only difference being that the relative lengths are upgraded to absolute lengths.

Euclidean geometry is simply metric geometry, but incorporates the correct scale of the image.

The scale can be fixed by knowing the dimensions of a certain object in the image [26].

In Euclidean geometry, second-order curves such as ellipses, parabolas and hyperbolas are easily defined.

In affine geometry it is possible to deal with ratios of vectors and barycenter's of points, but there is no way to express the notion of length of a line segment or to talk about orthogonality of vectors. A Euclidean structure allows us to deal with metric notions such as orthogonality and length (or distance) [94–96].

6.8 VANISHING POINTS AND LINES

A vanishing point (VP) is defined as Parallel lines of the world are projected into perspective images as intersecting lines and their point of intersection (Fig. 6.13). Each set of parallel lines is thus associated to a VP. Popular applications of VPs in computer vision are calibrated, rotation estimation and 3D reconstruction [97, 98].

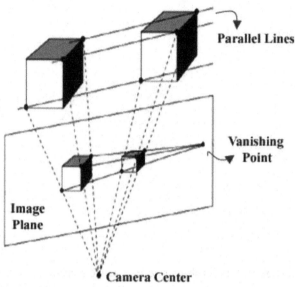

FIGURE 6.13   Overview of vanishing point.

The VP is the image of the point at infinity where the intersections of two or more parallel lines. it is computed as the least squares solution for the intersection of sets of images on parallel 3D line segments.

Vanishing lines are determined from vanishing points and parallelism constraints. Two or more vanishing points parallel to a plane define its vanishing line. A vanishing point belongs to the 3D direction perpendicular to a plane, completely defines the vanishing line [22].

Each vanishing point corresponds to an orientation in the 3D image and when the camera geometry is known, these orientations can be recovered. Even without this information, vanishing points can be used to group segments on the image with the same 3D orientation. Because of its important role in 3D reconstruction, the detection of the vanishing points in an image has to be effective, especially when no human intervention is required [32].

Estimating vanishing points in an image provides strong cues to make inferences about the 3D structures of a 2D image, such as depth and object dimension, because they are invariant features and they have many applications ranging from autonomous navigation to single-view reconstruction [98].

6.9   IMAGE FEATURES

In computer vision and image processing, a feature is a piece of information, which is relevant for solving the computational task related to a certain application. Features may be specific structures in the image such as points, edges or objects. Features may also be the result of a general neighborhood operation or feature detection applied to the image. Other examples of features are related to motion in image sequences, to shapes defined in terms of curves or boundaries between different image regions, or to properties of such a region. The feature concept is very general and the choice of features in a particular computer vision system may be highly dependent on the specific problem at hand [99–102].

The first kind of feature that you may notice are specific locations in the images, such as mountain peaks, building corners, doorways, or interestingly shaped patches of snow. These kinds of localized features are often called key point features or interest points (or even corners) and are often described by the appearance of patches of pixels surrounding the point location. Another class of important features is edges, for example, the profile of the mountains against the sky. These kinds of features can be matched based on their orientation and local appearance (edge profiles) and can also be good indicators of object boundaries and occlusion events in image sequences. Edges can be grouped into longer curves and straight line segments, which can be directly matched, or analyzed to find vanishing points and hence internal and external camera parameters.

6.9.1   EDGES

Edges are points where there is a boundary (or an edge) between two image regions. In general, an edge can be of almost arbitrary shape, and may include junctions. In practice, edges are usually defined as sets of points in the image, which have a strong gradient magnitude. Furthermore, some common algorithms will then chain high gradient points together to form a more complete description of an edge. These algorithms usually place some constraints on the properties of an edge, such as shape, smoothness, and gradient value. Locally, edges have a one-dimensional structure [103–105].

6.9.2   CORNERS/INTEREST POINTS

The terms corners and interest points are used somewhat interchangeably and refer to point-like features in an image, which have a local two-dimensional structure. The name "Corner" arose since early algorithms first performed edge detection, and then analyzed the edges to find rapid changes in direction (corners). These algorithms were then developed so that explicit edge detection was no longer required, for instance by looking for high levels of curvature in the image gradient. It was then noticed that the so-called corners were also being detected in parts of the image, which were not corners in the traditional sense (for instance a small bright spot on a dark background may be detected). These points are frequently known as interest points, but the term "corner" is used by tradition [104, 106].

6.9.3   BLOBS/REGIONS OF INTEREST OR INTEREST POINTS

A blob (alternately known as a binary large object, basic large object) is a collection of binary data stored as a single entity in a database management system. Blobs are typically images, audio or other multimedia objects, though sometimes-binary executable code is stored as a blob. Database support for blobs is not universal. Blobs provide a complementary description of image structures in terms of regions, as opposed to corners that are more point-like. Nevertheless, blob descriptors often contain a preferred point (a local maximum of an operator response or a center of gravity), which means that many blob detectors may also be regarded as interest point operators. Blob detectors can detect areas in an image, which are too smooth to be detected by a corner detector [107, 108].

6.10   FEATURE DETECTION

As noted, the first step in point correspondence is feature (or interest) point detection. Feature detection includes the use of gray level statistics and the detection of edges and corners. Methods based on detecting edges and corners are particularly useful in applications such as analysis of aerial images of urban images, airport facilities, and image to map matching, etc. Algorithms based on gray level statistics are applicable to a wider variety of images such as desert images and vegetation, which may or may not contain any man-made structures. Features, by definition, are locations in the image that are perceptually interesting. One can characterize an image feature detection algorithm by two attributes: (a) generality, and (b) robustness. Given that the nature of the salient features vary from application to application, it is desirable that a feature selection algorithm be as general as possible. In case of structured objects such features could be cornered and locations with significant curvature changes. When analyzing human faces, features of interest could be the eyes, nose, mouth, etc. [109].

The aim of key point detection is to determine points on a surface (a 3D face in our case), which can be identified with high repeatability in different range images of the same, surface in the presence of noise and pose variations. In addition to repeatability, the features extracted from these key points should be sufficiently distinctive in order to facilitate accurate matching. Methods based on image features: Such techniques extract features such as edges, corners, contours, and centroid of a specific region from the images and use the correlation between these features to determine the optimal alignment between the images. However, robustness cannot be guaranteed using only some sparse features. Human assistance is often needed in these algorithms, or the correct-match rate (CMT) will be relatively low [110, 111]. Table 6.4 is summarized some Common feature detectors and their classification.

TABLE 6.4 Common Feature Detectors and Their Classification

Feature detector	Edge	Corner	Blob
Canny	X		
Sobel	X		
Harris & Stephens/Plessey	X	X	
Difference of Gaussians		X	X
Hessian		X	X
MSER			X
SIFT			X
FORSTNER		X	

Successful image based object recognition methods recently developed are supported on the concept of local image descriptors. Image descriptors are localized information chunks extracted in particular points of the image that remain stable in face of common image transformations, and with the ability to distinguish between different patterns. Several types of local descriptors have been reported such as: Harris, Hessian, MSER, Forstner, SIFT, SURF, HOG, DTCWT, Gist, Ferns, Gepard, and Daisy.

In the framework above-mentioned, gradient based histograms showed the best performance. The initial steps consist in the interest point selection in scale space (e.g., Hessian, Harris), and the computation of the image gradients in the neighborhood of interest points (e.g., pixel differences, Canny detector). The descriptor is then obtained by splitting the interest point neighborhood into smaller regions, and finally for every subregion it is computed the histogram of the gradient orientation with an appropriate information selection procedure (e.g., weighting, Principal Component Analysis-PCA). To date, the most remarkable descriptor in terms of distinctiveness is the SIFT local descriptor, which computes the image gradient from pixel differences, subdivide the interest point regions in a Cartesian grid, and for each subregion, compute the gradient orientation histogram weighted by the gradient magnitude [112, 113].

6.10.1 HARRIS

The Harris detector is one of most widely used detectors for finding feature points. It is based on detecting changes in image intensity around a point using the auto correlation matrix where the matrix is composed of first order image derivatives. Originally, small filters were used to calculate the image derivatives. The most popular interest point operators are the Harris corner detector and the Good Features to Track, also referred to as Shi-Tomasi features. It defines key points to be "points that have locally maximal self-matching precision under translational least-squares template matching." The Harris affine detector relies on interest points detected at multiple scales using the Harris corner measure on the second-moment matrix. In practice, these key points often correspond to corner-like structures.

Harris detector computes a matrix related to the autocorrelation function of the image. The squared first derivatives of the image signal are averaged over a window and the eigenvalues of the resulting matrix are the principal curvatures of the auto-correlation function. An interest point is detected if the found two curvatures are high. Harris points are invariant to rotation [50, 114, 115].

The Harris detector proceeds by searching for points x where the second-moment matrix C around x has two large eigenvalues. The matrix C can be computed from the first derivatives in a window around x, weighted by a Gaussian $G(x,\tilde{\sigma})$:

$$C(x,\sigma,\tilde{\sigma}) = G(x,\tilde{\sigma}) * \begin{bmatrix} I_x^2(x,\sigma) & I_x I_y(x,\sigma) \\ I_x I_y(x,\sigma) & I_y^2(x,\sigma) \end{bmatrix} \tag{3}$$

In this formulation, the Gaussian $G(x,\tilde{\sigma})$ takes the role of summing over all pixels in a circular local neighborhood, where each pixel's contribution is additionally weighted by its proximity to the center point. Instead of explicitly computing the eigenvalues of C, the following equivalences (4) are used:

$$\det(C) = \lambda_1 \lambda_2$$
$$trace(C) = \lambda_1 + \lambda_2 \tag{4}$$

to check if their ratio $r = \dfrac{\lambda_1}{\lambda_2}$ is below a certain threshold. With

$$\frac{trace^2(C)}{\det(C)} = \frac{(\lambda_1 + \lambda_2)^2}{\lambda_1 \lambda_2} = \frac{(r\lambda_2 + \lambda_2)^2}{r\lambda_2^2} = \frac{(r+1)^2}{r} \tag{5}$$

this can be expressed by the following condition

$$\det(C) - \alpha trace^2(C) \rangle t \tag{6}$$

which avoids the need to compute the exact eigenvalues. Typical values for α are in the range of 0.04–0.06. The parameter $\tilde{\sigma}$ is usually set to 2σ, so that the considered image neighborhood is slightly larger than the support of the derivative operator used The detection procedure is visualized in Table 6.5 [115].

TABLE 6.5   Harris Detector

Future	Corner	Edge	Flat
Harris characteristics	$\lambda_1 \sim \lambda_2$	$\lambda_1 \ll \lambda_2$	λ_1, λ_2 are small

6.10.2 HESSIAN

The Hessian affine region detector is a feature detector used in the fields of computer vision and image analysis. Like other feature detectors, the Hessian affine detector is typically used as a preprocessing step to algorithms that rely on identifiable, characteristic interest points. The Hessian affine detector is part of the subclass of feature detectors known as affine-invariant detectors: Harris affine region detector, Hessian affine regions, maximally stable extremal regions, Kadir–Brady saliency detector, edge-based regions (EBR) and intensity-extrema-based (IBR) regions [116].

The detector uses the Hessian matrix to locate points in the image plane and a Laplacian function to compute scale for those points. A localization step ensures that the location and scale of the points detected is close to their true location. The Hessian affine also uses a multiple scale iterative algorithm to spatially localize and select scale & affine invariant points. However, at each individual scale, the Hessian affine detector chooses interest points based on the Hessian matrix at that point. Like the Harris affine algorithm, these interest points based on the Hessian matrix are also spatially localized using an iterative search based on the Laplacian of Gaussians. Predictably, these interest points are called Hessian–Laplace interest points. Furthermore, using these initially detected points, the Hessian affine detector uses an iterative shape adaptation algorithm to compute the local affine transformation for each interest point. The implementation of this algorithm is almost identical to that of the Harris affine detector; however, the above-mentioned Hessian measure replaces all instances of the Harris corner measure.

The Hessian matrix as derived from the second order Taylor series expansion can be used to describe the local structure around a point. For an image I, the Hessian matrix can be expressed as:

$$H = \begin{bmatrix} I_{xx} & I_{xy} \\ I_{yx} & I_{yy} \end{bmatrix} \tag{7}$$

$$DET = Det(H) = I_{xx}I_{yy} - I_{xy}^2 \tag{8}$$

where I_{xx}, I_{yy} and I_{xy} are the second order derivatives of image intensity. The extrema of the DET measure in a local neighborhood was used to detect interest points [116]. Hessian detector calculates the corner strength as the determinant of the Hessian matrix $(I_{xx}I_{yy} - I_{xy}^2)$. The local maxima of the corner strength denote the corners in the image. The determinant is related to the Gaussian curvature of the signal and this measure is invariant to rotation. An extended version, called Hessian- Laplace detects points, which are invariant to rotation and scale (local maxima of the Laplacian-of-Gaussian) [117].

In general, it can be stated that Harris locations are more specific to corners, while the Hessian detector also returns many responses on regions with strong texture variation. In addition, Harris points are typically more precisely located as a result of using first derivatives rather than second derivatives and of taking into account a larger image neighborhood. Thus, Harris points are preferable when looking for exact corners or when precise localization is required, whereas Hessian points can provide additional locations of interest that result in a denser cover of the object [118].

6.10.3 MSER

In computer vision, maximally stable extremal regions (MSER) are used as a method of blob detection in images. This technique finds correspondences between image elements from two images with different viewpoints. This method extracts regions closed under continuous transformation of the image coordinates and under monotonic transformation of the image intensities to the wide-baseline matching and it has led to better stereo matching and object recognition algorithms. In this case, the features are actually shapes rather than points or corners. This detector can be described in simple terms using its similarity to the watershed algorithm based on the actual intensity values found in the image, but it is also possible to describe MSERs by their shape alone. When combined with certain descriptors, MSERs perform very well when detected on flat surfaces, and have average performance for use with images of 3D objects. They can also work well for changes in illumination between images [117,119].

MSERs are regions that are either darker, or brighter than their surroundings, and that are stable across a range of thresholds of the intensity function. MSERs have also been defined on other scalar functions, and have been extended to color [120].

The concept more simply can be explained by thresholding. All the pixels below a given threshold are 'black' and all those above or equal are 'white.' In the case of MSERs, the total number of regions is taken to be twice the number of MSERs that match to account for the fact that there are two patches for each region, one shape patch, and one texture patch and thus two possible matches. The second graph, 1-precision, is given by

$$1 - precision = \frac{\#\text{false matches}}{\#\text{correct matches} + \#\text{false matches}} \quad (9)$$

where a lower number is better. The last graph depicts what percentage of all matches were found to be correct [119].

6.10.4 FORSTNER

Forstner detector uses also the autocorrelation function to classify the pixels into categories (interest points, edges or region); the detection and localization stages are separated, into the selection of windows, in which features are known to reside, and feature location within selected windows. Further statistics performed locally allow estimating automatically the thresholds for the classification. The algorithm requires a complicate implementation and is generally slower compared to other detectors [117].

Both the Harris detector and the Forstner operator are determined with similar concept of differential calculation. They are used to determine the magnitude and the direction of the intensity value changes in an image. A continuous filter function is approximated by the first derivative of the Gaussian. The derivative of the smoothed image corresponds to a convolution of the image with the derivative of the Gaussian. The convolution and differentiation are linear shift-invariant operations. Given a two-dimensional image function $f(x, y)$, the gradient

$$\nabla f = (f_x, f_y)^T \tag{10}$$

is defined by the two dimension spatial derivatives of f in x- and y-direction

$$f_x = f * G_x, f_y = f * G_y \tag{11}$$

where * denotes the discrete convolution

$$f(x,y) * g(x,y) = \sum_{i=-r}^{r} \sum_{j=-r}^{r} f(x-i, y-j) \cdot g(i,j) \tag{12}$$

of the image f with a filter kernel g and

$$G_x(x,y) = \frac{\partial G_\sigma(x,y)}{\partial x} = -\frac{x}{2\pi\sigma^4} \cdot \exp\left(-\frac{x^2 + y^2}{2\sigma^2}\right), \tag{13}$$

$$G_y(x,y) = \frac{\partial G_\sigma(x,y)}{\partial y} = -\frac{y}{2\pi\sigma^4} \cdot \exp\left(-\frac{x^2 + y^2}{2\sigma^2}\right) \tag{14}$$

Denotes spatial derivatives in x- and y-direction of a Gaussian

$$G_\sigma(x,y) = \frac{1}{2\pi\sigma^4} \cdot \exp\left(-\frac{x^2 + y^2}{2\sigma^2}\right) \tag{15}$$

Forstner also identifies salient points by the use of autocorrelation matrix A. First, the derivatives are computed on a smoothed image with the natural scale σ, and are then summed over a Gaussian window using an artificial scale σ_2 with the structure tensor:

$$A = G_{\sigma_2} \cdot \begin{bmatrix} f_x^2 & f_x f_y \\ f_x f_y & f_x^2 \end{bmatrix} \tag{16}$$

Where the indices of the sums over the area Ω were omitted for simplicity [121,122].

6.10.5   SIFT

The SIFT (Scale Invariant Feature Transform) has been shown to perform better than other local descriptors. Given a feature point, the SIFT descriptor computes the gradient vector for each pixel in the feature point's neighborhood and builds a normalized histogram of gradient directions. The descriptor computation starts from a scale and rotation normalized region extracted with one of the above-mentioned detectors. As a first step, the image gradient magnitude and orientation are sampled around the key point location using the region scale to select the level of Gaussian blurs (i.e., the level of the Gaussian pyramid at which this computation is performed). The SIFT descriptor creates a 16×16 neighborhood that is partitioned into 16 subregions of 4×4 pixels each with 8 orientation bins each, weighted by the corresponding pixel's gradient magnitude. Once all orientation histogram entries have been completed, those entries are concatenated to form a single $4 \times 4 \times 8 = 128$ dimensional feature vector [123, 124].

This descriptor aims to achieve robustness to lighting variations and small positional shifts by encoding the image information in a localized set of gradient orientation histograms. The purpose of this Gaussian window is to give higher weights to pixels closer to the middle of the region, which is less affected by positional shifts. At the same time, the high-dimensional representation provides enough discriminative power to reliably distinguish a large number of key points. When computing the descriptor, it is important to avoid all boundary effects, both with respect to spatial shifts and to small orientation changes. Thus, when entering a sampled pixel's gradient information into the 3D spatial/ orientation histogram, its contribution should be smoothly distributed among the adjoining histogram bins using trilinear interpolation. Final illumination normalization completes the extraction procedure. For this, the vector is first normalized to unit length, thus adjusting for changing image contrast. Then all feature dimensions are threshold to a maximum value of 0.2 and the vector is again normalized to unit length [124–127].

The scale-invariant features are efficiently identified by using four major staged filtering approaches: (1) scale-space peak selection; (2) key point localization; (3) orientation assignment; (4) key point descriptor. In the first stage, potential interest points are identified by scanning the image over location and scale. This is implemented efficiently by constructing a Gaussian pyramid and searching for local peaks (termed key points) in a series of difference-of-Gaussian (DoG) images. In the second stage, candidate key points are localized to subpixel accuracy and eliminated if found to be unstable. The third identifies the dominant orientations for each key point based on its local image patch. The assigned orientation(s), scale and location for each key point enables SIFT to construct a canonical view for the key point that is invariant to similarity transforms.

The final stage builds a representation descriptor for each key point, based upon the image gradients in its local neighborhood (discussed below in greater detail). Note that the image gradients has been previously centered about the key point's location, rotated on the basis of its dominant orientation and scaled to the appropriate size. The goal is to create a descriptor for the key point that is compact, highly distinctive and yet robust to changes in illumination and camera viewpoint (i.e., the same key point in different images maps to similar representations) [128, 129].

In order to compute the local descriptor, the regions are scale normalized to compute the derivatives I_x and I_y of the image I with pixel differences:

$$I_x(x,y) = I(x+1,y) - I(x-1,y) \tag{17}$$

$$I_y(x,y) = I(x,y+1) - I(x,y-1) \tag{18}$$

Image gradient magnitude and orientation is computed for every pixel in the image region:

$$M(x,y) = \sqrt{I_x(x,y)^2 - I_y(x,y)^2} \tag{19}$$

$$\Theta(x,y) = \tan^{-1}\left(\frac{I_y(x,y)}{I_x(x,y)} \right) \tag{20}$$

The interest region is then divided in subregions in a rectangular grid. The next step is to compute the histogram of gradient orientation, weighted by gradient magnitude, for each subregion. Orientation is divided into

8 bins and each bin is set with the sum of the windowed orientation difference to the bin center, weighted by the gradient magnitude:

$$h_{r(l,m)}(k) = \sum_{x,y \in r(l,m)} M(x,y)\left(1 - \left|\Theta(x,y) - c_k\right|\Big/\Delta_k\right)$$

$$\Theta(x,y) \in bin \quad k \tag{21}$$

Where c_k is the orientation bin center, Δ_k is the orientation bin width, and (x, y) are pixel coordinates in subregion $r(l,m)$.

The SIFT local descriptor is the concatenation of the several gradient orientation histograms for all subregions:

$$u = \left(h_{r(1,1)},...,h_{r(l,m)},...,h_{r(4,4)}\right) \tag{22}$$

The final step is to normalize the descriptor in Eq. (22) to unit norm, in order to reduce the effects of uniform illumination changes. The gradient orientation is not invariant to rotations of the image region, so the descriptor is not invariant. To provide orientation invariance, Lowe proposed to compute the orientation of the image region, and set the gradient orientation relative to the region's orientation. The orientation is given by the highest peak of the gradient orientation histogram of the image region. In further object recognition tests, we compute both invariant and noninvariant descriptors [113, 124].

The SIFT keys derived from an image are used in a nearest-neighbor approach to indexing to identify candidate object models. Collections of keys that agree on a potential model pose are first identified through a Hough transform hash table, and then through a least-squares fit to a final estimate of model parameters. When at least three keys agree on the model parameters with low residual, there is strong evidence for the presence of the object. Since there may be dozens of SIFT keys in the image of a typical object, it is possible to have substantial levels of occlusion in the image and yet retain high levels of reliability (Fig. 6.14).

The current object models are represented as 2D locations of SIFT keys that can undergo affine projection. Sufficient variation in feature location is allowed to recognize perspective projection of planar shapes at up to a 60-degree rotation away from the camera or to allow up to a 20-degree

FIGURE 6.14   Hough transform example (simplified).

rotation of a 3D object. SIFT was mainly developed for gray images which limits its performance with some colored objects (Fig. 6.15) [129, 130].

6.10.6 SURF

SURF (Speeded up Robust Features) is one of the best interest point detectors and descriptors currently available. It has been shown to outperform the other well-known methods based on interest points SIFT and GLOH (Gradient Location and Orientation Histogram).

The SURF technique uses a Hessian matrix based blob detector to find interest points and a sum of 2D Haar wavelet responses within the interest point neighborhood as descriptor, calculated in a 4×4 subregion around each interest point. Interest points are localized in scale and image space by applying non-maximum suppression in a 3×3×3 neighborhood.

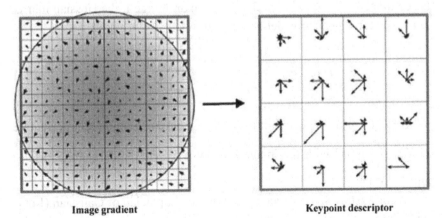

Image gradient   Keypoint descriptor

FIGURE 6.15   The SIFT descriptor for a feature point [125].

When the dominant orientation is estimated and included in the interest point information, SURF descriptors are constructed by extracting square regions around the interest points. An image is analyzed at several scales, so interest points can be extracted from both global ('coarse') and local ('fine') image details. Additionally, the dominant orientation of each of the interest points is determined to support rotation-invariant matching [131].

The standard SURF descriptor has a dimension of 64 and the extended version (e-SURF) of 128. The u-SURF version is not invariant to rotation and has a dimension of 64. Moreover, due to reducing the time for feature computation and matching, and increasing simultaneously the robustness, only 64 dimensions are used [132].

The proposed SURF descriptor is based on similar properties, with a complexity stripped down even further. The first step consists of fixing a reproducible orientation based on information from a circular region around the interest point. Then, we construct a square region aligned to the selected orientation, and extract the SURF descriptor from it [133].

One of the main advantages of SURF is to be able to compute distinctive descriptors quickly. In addition, SURF descriptor is invariant to common image transformations including image rotation, scale changes, illumination changes, and small change in viewpoint.

The detection process in this technique, is based on the Hessian matrix, but uses a very basic approximation, just as DoG is a very basic Laplacian-based detector. Given a point $X = (x, y)$ in an image I, the Hessian matrix $H(X, \sigma)$ at X at scale σ is defined as follows:

$$H(X,\sigma) = \begin{bmatrix} L_{xx}(X,\sigma) & L_{xy}(X,\sigma) \\ L_{xy}(X,\sigma) & L_{yy}(X,\sigma) \end{bmatrix} \tag{23}$$

Where $L_{xx}(X, \sigma)$ is the convolution of the Gaussian second order derivative $I(X) * \dfrac{\partial^2}{\partial x^2} g(\sigma)$ with the image I at point X, and similarly for $L_{xy}(X, \sigma)$ and $L_{yy}(X, \sigma)$.

In contrast to SIFT, which approximates Laplacian of Gaussian (LoG) with Difference of Gaussians (DoG), SURF approximates second order

Gaussian derivatives with box filters (mean or average filter). These can be calculated rapidly through integral images [134–136].

6.10.7 HOG

The Histogram of Gradient Orientation (HoG) descriptor is widely applied in object classification, and human (e.g., pedestrian) detection. Locally normalized HOG descriptors provide excellent performance relative to other existing feature sets including wavelets. The HoG descriptor has gained traction in the vision community, and particularly in object recognition, for several reasons. First, it is a vector-space model, where perceptual similarity is approximated by Euclidian (or cosine) distance between two HoG vectors. This means that many off-the-shelf learning and database algorithms can work directly on HoG representations. Second, it appears to be a reasonably good model of perceptual similarity: it uses intensity gradients rather than intensity directly, which means that the responses of edges are localized; it is sensitive to local but not global contrast due to its normalization scheme; it can handle minor misalignment due to the bilinear interpolation between HoG cells; and many other reasons also apply. Third, it is very fast to compute: computing a HoG pyramid for a 500-by-500 image can take less than 2 seconds on a single core, and firing a sliding-window template at all positions and scales can happen equally fast via fast Fourier- transform convolution [137,138].

The standard HoG algorithm begins by computing a gradient image in each of 9 orientations. Then at each pixel, it finds the gradient image with the largest gradient magnitude at that pixel, and then add the gradient's magnitude to the histograms corresponding to the maximal orientation [138].

To enable histogram construction, the range of edge orientations is quantized into q bins. The histogram counts are concatenated to form a q-D vector for each cell, which are again concatenated to form an qn^2-D vector for the window. In many implementations, several windows are sampled in a non-overlapping w × w grid local to the key-point and again concatenated to output the final descriptor. The aim of such method is to describe an image by a set of local histograms. These histograms count occurrences of gradient orientation in a local part of the image [139,140].

6.10.8 DTCWT

Dual-tree complex wavelet transform (DTCWT) was shown to be a particularly suitable tool for image analysis as it is directionally selective, approximately shift invariant, limited redundancy, and computing efficiency.

That is, any finite energy analog signal $E(C)$ can be decomposed in terms of wavelets and scaling functions via:

$$E(C) = \alpha^s \left(\prod_{b=1}^{6} \rho_b \right)^\beta \tag{24}$$

where ρ_b is the DTCWT coefficients and Parameters α & β are the scaling coefficients and wavelet coefficients. They are computed via the inner products and control the relative weight of scales in the accumulated map. They provide a time-frequency analysis of the signal by measuring its frequency content (controlled by the scale factor α) at different times (controlled by the time shifts).

DTCWT is implemented by two discrete wavelet trees to compute real part and imaginary part. They work on the row and column of image [141].

The DTCWT is approximately shift invariant and offers lower redundancy with greater computational efficiency. The DTCWT descriptor is more efficient than the existing popular scale and rotation invariant method of SIFT. It is adapted here to provide image matching between a small template and a larger image rather than matching key points of two similarly sized images [142].

In multi resolution analysis of wavelets, the low pass information consists of approximated version of high resolution image. The high pass information gives sharper variation details. In dual tree complex wavelets, two trees are used in parallel to generate the output interpreting them as real and imaginary part of complex coefficient. Two sets of QMF (quadrature mirror filter) filter pairs are used in the generation of real and imaginary coefficients in the analysis branch. The transform decomposes the image into sub bands providing the two smoothed versions of image and the information in six different directions at each stage [143–145].

6.10.9   GIST

The GIST has been shown effective to suitably model semantically meaningful and visually similar scenes and has been used effectively for retrieving nearest neighbors from large scale image databases. Although the descriptor is not very discriminative, the attractive feature of this representation is that it is very compact, fast to compute and that roughly encodes spatial information. The GIST descriptor enables us to obtain a small number of visually similar clusters of alike locations, for example, large open areas, suburban areas, large high-rise buildings, etc. [146, 147].

The GIST descriptor is a global descriptor of an image that represents the dominant spatial structures of the scene in the image. Each image is represented by a 320 dimensional vector (per color band). The feature vector corresponds to the mean response to steerable filters at different scales and orientations computed over 4×4 sub-windows. For a more intuitive idea of what this descriptor encloses, see the clustering results and average images obtained when clustering gist values from a big set of outdoor images. The GIST descriptor aggregates oriented edge responses at multiple scales into very coarse spatial bins. The advantage of this descriptor is that it is very compact and fast to compute [146, 148, 149].

Generally the GIST descriptor focuses on the shape of scene itself, on the relationship between the outlines of the surfaces and their properties, and ignores the local objects in the scene and their relationships. The representation of the structure of the scene, termed spatial envelope is defined, as well as its five perceptual properties: naturalness, openness, roughness, expansion and ruggedness, which are meaningful to human observers. The degrees of those properties can be examined using various techniques, such as Fourier transform and PCA. The contribution of spectral components at different spatial locations to spatial envelope properties is described with a function called windowed discriminant spectral template (WDST), and its parameters are obtained during learning phase [150].

One example of a global image representation is the GIST descriptor . The GIST descriptor is a vector of features g, where each individual feature g_k is computed as:

$$g_k = \sum_{x,y} w_k(x,y) \times |I(x,y) \otimes h_k(x,y)|^2 \qquad (25)$$

where $\otimes$ denotes image convolution and $\times$ is a pixel-wise multiplication. $I(x, y)$ is the luminance channel of the input image, $h_k(x, y)$ is a filter from a bank of multiscale oriented

Gabor filters (6 orientations and 4 scales), and $w_k(x, y)$is a spatial window that will compute the average output energy of each filter at different image locations. The windows $w_k(x, y)$ divide the image in a grid of 4×4 non-overlapping windows. This results in a descriptor with a dimensionality of $4 \times 4 \times 6 \times 4 = 384$ [151, 152].

6.10.10   FERNS

The KD-Ferns, a novel algorithm for fast approximate nearest neighbors enabling a reduction in the number of searched parts in each image location. To speed up the coarse level the KD-Ferns algorithm is used to compare only a small subset of the parts to each image location. In addition, for a specific part, only a sparse set of locations is considered using spatial inhibition [153].

The original Ferns approach uses an extrema of Laplacian operator to detect interest points in input images and stores probabilities as 4-byte floating point values. This was replaced by the FAST detector with non-maximum suppression on two octaves of the image. At runtime, the FAST threshold is dynamically adjusted to yield a constant number of interest points (300 for a 320×240 input image). FAST typically shows multiple responses for interest points detected with more sophisticated methods. It also does not allow for subpixel accurate or scale-space localization. These deficiencies are counteracted by modifying the training Scheme to use all FAST responses within the 8-neighborhood of the model point as training examples. Except for this modification, the training phase (running on the PC) is performed exactly as described in Ref. [153].

Ferns, unlike randomized trees, are nonhierarchical structures. Each fern consists of a set of ordered binary tests and returns the probability that a patch belongs to each of the classes learnt during training. The structure of ferns is shown in Fig. 6.16. Each node in a fern returns a binary digit based on the result of the test. Hence, a fern with S nodes will return a number between 0 and 2S − 1. For multiple patches that belong to the same class, the output of a fern for that class can be modeled with a multinomial

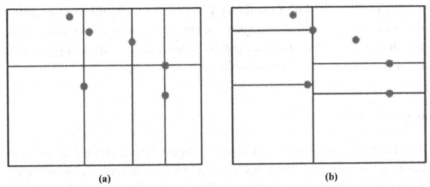

FIGURE 6.16   Space partition for 6 points in 2D using (a) the KD-Fern (b) and the KD-Tree.

distribution. During classification, performing the set of tests on a test patch also returns a binary code, which is used to obtain the likelihood of occurrence in each class. The class is found by multiplying class probabilities across the ferns, and selecting the class that gives the maximum product. Performance-memory trade-offs can be made by changing the size and number of ferns allowing for flexible implementation.

A "KD-Fern" is a KD-Tree with the following property: all nodes in the same level (depth) of the tree have the same splitting dimension d and threshold τ. The search algorithm is identical to the one described for the KD-Tree but due to its restricted form can be implemented more efficiently [154–158].

6.10.11   GPARD

Gepard performs slightly better even with small number of principal components, due to the final ESM (efficient second-order minimization method) refinement [159].

Instead of computing the naive high dimensional descriptor, Gepard compute a much lower dimensional descriptor whose elements are made of mean patches each of them computed for a small range of local poses.

In this descriptor, the mean patches can be computed much more efficient which concept of the mean patch is similar to the "Geometric Blur." One characteristic of a mean patch is that its center is less blurred than its outer parts. Respond of a similarity function between an incoming patch and a mean patch is high enough to be considered as a valid match, if the transformation of the incoming patch comes from the range of poses that the mean patch was computed with.

6.10.12   DAISY

The DAISY descriptors cover a wide range of descriptors, because it is easily reconfigurable. The optimization is based on three outdoor scenes where ground truth is obtained from the bundler software, which is based on the SIFT framework. DAISY is much faster to compute than SIFT, which allows its descriptors to be calculated densely for every image pixel, enabling wide baseline stereo matching. Additionally, the DAISY setup allows to make the descriptor rotation invariant by rotating the descriptor itself, instead of the full input patch, which requires less control signals and therefore simplifies the design. It should be noted, though, that the descriptor can only be rotated to discrete orientations [160,161].

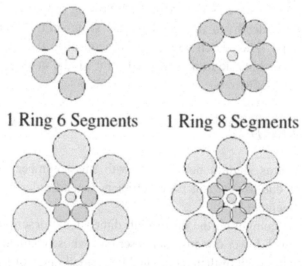

1 Ring 6 Segments   1 Ring 8 Segments

2 Rings 6 Segments   2 Rings 8 Segments

FIGURE 6.17   Typical DAISY descriptor Gaussian summation regions (circles indicate 1 standard deviation) [161].

DAISY samples local gradient information in the way visualized by Fig. 6.17. Each circle represents one histogram region, which is part of the descriptor vector. Each histogram represents the gradient orientations within this region.

The original DAISY algorithm already calculates values very similar to SIFT's main orientation histogram. DAISY usually discretizes the gradient orientations into 8 directions, which means that every histogram bin corresponds to 45° of the full 360°. The gradient magnitude is calculated for each discrete direction, resulting in one gradient image per direction, the so called orientation maps. Each orientation map is now sequentially smoothed with Gaussian masks, resulting in eight smoothed orientation maps, one for each orientation. The magnitudes of the smoothed gradients are the entries that will be written into the final descriptor [162].

6.11   FEATURE MATCHING

The matching strategy is based on a similarity measure that reflects the positioning of features extracted through the first-order derivative operators, and which also quantifies the contribution of any additional attribute, which can be associated with these features [109, 163].

The goal of correspondence estimation is to take a raw set of images and to find sets of matching 2D pixels across all the images. Each set of matching pixels, ideally represents a single point in 3D [30].

The above features are sufficiently well characterized to use a very simple one-shot matching scheme. Each feature is represented by a set of affine invariants, which are combined into a feature vector. A Mahalanobis distance metric is used to measure similarity between any two feature vectors. Hence for two images we have a set of feature vectors $v^{(i)}$ and $w^{(i)}$ and a distance measure $d(v, w)$. The matching scheme proceeds as follows:

1. calculate distance matrix m_{ij} between pairs of features across the two images.

$$m_{ij} = d\left(v^{(i)}, w^{(i)}\right) \qquad (26)$$

2. identify potential matches (i, j) such that the feature i is the closest feature in the first image to feature j in the second image and vice versa.

3. score matches using an ambiguity measure.
4. select unambiguous matches (or best "n" matches) [164].

An alternative to tracking features from one frame to another is to detect features independently in each image, and then search for the corresponding features on the basis of a "matching score" that measure how likely it is for two features to correspond to the same point in space. The generally accepted technique consists in first establishing putative correspondences among a small number of features, and then trying to extend the matching to additional features whose score falls within a given threshold by applying robust statistical techniques [86].

Feature extractors such as SIFT take an image and return a set of pixel locations that are highly distinctive. For each of these features, the detector also computes a "signature" for the neighborhood of that feature, also known as a feature descriptor: a vector describing the local image appearance around the location of that feature. Once the features have been extracted from each image, then features match by finding similar features in other images [30].

For example, in the study of 3D object recognition and reconstruction by M. Brown et al. [165], SIFT (Scale Invariant Feature Transform) has been used for feature matching. These locate interest points at maxima/minima of a difference of Gaussian function in scale-space. Each interest point has an associated orientation, which is the peak of a histogram of local orientations. This gives a similarity invariant frame in which a descriptor vector is sampled. Though a simple pixel resampling would be similarity invariant, the descriptor vector actually consists of spatially accumulated gradient measurements. This spatial accumulation is important for shift invariance, since the interest point locations are typically accurate in the 1–3 pixel range [26]. Illumination invariance is achieved by using gradients (which eliminates bias) and normalizing the descriptor vector (which eliminates gain). Once features have been extracted from all n images (linear time), they must be matched. Since multiple images may view the same point in the world, each feature is matched to k nearest neighbors (typically k = 4). This can be done in O ($nlog\ n$) time by using a k-d tree to find approximate nearest neighbors [165].

However, in the past ten years, more powerful feature extractors have been developed that achieve invariance to a wide class of image transformations, including rotations, scales, changes in brightness or contrast, and, to some extent, changes in viewpoint (Fig. 6.18). These techniques allow us to match features between images taken with different cameras, with different zoom and exposure settings, from different angles, and V in some cases V at completely different times of day. Thus, recent feature extractors open up the possibility of matching features in Internet collections, which vary along all of these dimensions (and more) [30].

The performance of the matching algorithm at a particular threshold can be quantified by first counting the number of true and false matches and match failures, using the following definitions:

- *TP*: true positives, i.e., number of correct matches;
- *FN*: false negatives, matches that were not correctly detected;
- *FP*: false positives, estimated matches that are incorrect;
- *TN*: true negatives, non-matches that were correctly rejected.
- True positive rate *TPR*,

$$TRP = \frac{TP}{TP + FN} = \frac{TP}{P} \tag{27}$$

- False positive rate *FPR*,

$$FPR = \frac{FP}{FP + TN} = \frac{FP}{N} \tag{28}$$

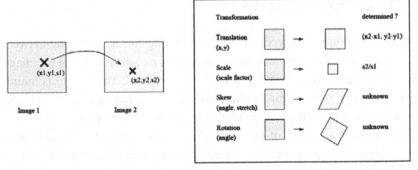

FIGURE 6.18 Hypothesized correspondence [164].

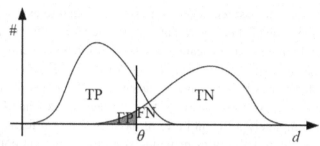

FIGURE 6.19   The distribution of positives (matches) and negatives (non-matches) as a function of inter-feature distance d. As the threshold is increased, the number of true positives (TP) and false positives (FP) increases.

Any particular matching strategy (at a particular threshold or parameter setting) can be rated by the TPR and FPR numbers: ideally, the true positive rate will be close to 1, and the false positive rate close to 0. Figure 6.1 shows how we can plot the number of matches and non-matches as a function of inter-feature distance d.

The problem with using a fixed threshold is that it is difficult to set; the useful range of thresholds can vary a lot as we move to different parts of the feature space a better strategy in such cases is to simply match the nearest neighbor in feature space. Since some features may have no matches (e.g., they may be part of background clutter in object recognition, or they may be occluded in the other image), a threshold is still used to reduce the number of false positives.

6.12   KEYPOINT MATCHING

There are mainly two types of image matching forms, matching between two images of the same size and searching for a subarea out of an image, which is most similar with the predefined template. The former is mainly used in image retrieval and pattern recognition. Finding correspondences between feature points is one of the keystones of computer vision, with application to a variety of problems. Automatic feature matching is often an initialization procedure for more complex tasks, such as fundamental matrix estimation, image mosaicking, object recognition, and three-dimensional point clouds registration [31, 44, 166].

Classical approaches to point matching with unknown geometry assume a short baseline, and they are usually based on correlation. It is well known that correlation based approaches suffer from view-point changes and do not take into account the global structure of the image. Recently, it has been applied to a variety of computer vision and pattern matching problems, including point and shapes matching, and image segmentation. The point matching method is based on computing a proximity matrix that depends on the distance between points belonging to the two images. The method performs well on synthetic images, but it is sensitive to the noise that affects point's detection and localization in real images [4, 35, 37, 43, 45, 167].

6.12.1 SHAPE-BASED MATCHING

A satisfactory theory of shape representation would have a number of desirable attributes:

1. it should support recognition based on exquisitely fine differences e.g. distinguishing faces of twins.
2. at the same time, it should support making coarse discriminations very quickly.
3. the approach should scale to deal with a large number of objects.
4. it should be possible to acquire a representation of an object category from relatively few examples, that is, there should be good generalization ability [168].

A shape is represented by a discrete set of points sampled from the internal or external contours on the shape. These can be obtained as locations of edge pixels as found by an edge detector, giving us a set of n points. Consider the set of vectors originating from a point to all other sample points on a shape. These $n - 1$ vectors express the configuration of the entire shape relative to the reference point. One way to capture this information is as the distribution of the relative positions of the remaining $n - 1$ points in a spatial histogram. Concretely, for a point p_i on the shape, compute a coarse histogram h_i of the relative coordinates of the remaining $n - 1$ point,

$$h_i(k) = \#\{q \neq p_i : (q - p_i) \in bin(k)\} \tag{29}$$

This histogram is defined to be the shape context of p_i.

There is a natural way to measure the similarity between two shape contexts. Consider a point p_i on the first shape and a point q_j on the second shape. Let $C_{ij} = C(p_i, q_j)$ denote the cost of matching these two points. As shape contexts are distributions represented as histograms, it is natural to use the distance:

$$C_{ij} = \frac{1}{2} \sum_{k=1}^{K} \frac{\left[h_i(k) - h_j(k) \right]^2}{h_i(k) + h_j(k)} \tag{30}$$

where $h_i(k)$ and $h_j(k)$ denote the K-bin normalized histogram at p_i and q_j, respectively [168].

6.12.2 BLOB MATCHING

The final step in the tracking part is the blob matching where the previous frame blobs are matched against the new frame blobs using the knowledge of the high level occlusions status. The matching is done using the distance between the previous objects centers and the new objects centers (Fig. 6.20) [169].

Model based blob tracking is one of the most important tracking methods and the similarity measure employed to measure the difference between the target model and target candidates is a very crucial issue influencing the tracking performance. The most often used similarity measures include Bhattacharyya coefficient, cross correlation, mean square difference, histogram intersection distance and so on [170].

In blob tracking task, the matching precision is quite important. A blob matching model is developed to evaluate the matching performance of different similarity measures numerically. The similarity measure employed is the most crucial one influencing the localization precision. In order to achieve satisfactory tracking performance, the blob matching is modeled as a maximization problem of a function with matching position parameter. While the matching performance of different similarity measures now can only be tested by visual experiments, thus the matching precision cannot be predicted or analyzed for certain image sequence in theory.

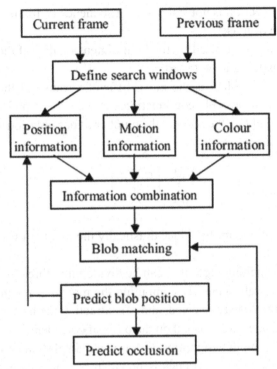

FIGURE 6.20   System architecture of blob matching.

The Bhattacharyya coefficient matching performance for visual tracking is analyzed in detail based on this model. Visual experiments have validated the matching model. Other similarity measures can be analyzed by this model also. Based on the analysis of Bhattacharyya coefficient employing the blob matching model, a modified version of Bhattacharyya coefficient is proposed. This modified Bhattacharyya coefficient has altered the power comparing with original Bhattacharyya coefficient and can obtain higher matching accuracy [170].

A blob-matching procedure is used for tracking blobs. Blob matching is a difficult task owing to the many problems involved:

1. Many blobs may appear or vanish in successive frames because of shadows, highlights, and other events.
2. Blobs may be occluded by the overlapping of other blobs.

3. Blob detection is not stable over a sequence of frames, as a consequence of noise.

4. Problems are caused by the high dimensionality of the set of possible matches to be tested.

5. The last problem is the most difficult because of the very large number of possible combinations. If we have α blobs in the first image and β in the second, the number of possible correspondences is given by:

$$\sum_{i=1}^{\min(\alpha,\beta)} \binom{\alpha}{i}\binom{\beta}{i} \tag{31}$$

It is therefore necessary to reduce drastically the number of combinations to be tested.

The blob-matching algorithm can be divided into three main steps: (a) reducing the search range, (b) computing the matching function for each pair of blobs previously selected, and (c) extracting the best matches.

Blob candidates are selected on the basis of considerations about some constraints. The main assumption is that if the displacement of an object between two consecutive frames is bounded by some constraints, there also exists a bound to the displacement of the corresponding blobs on the image plane (motion constraint). This constraint relies on the assumption that a blob is not expected to be too far from where it was in the previous frame. This assumption is reasonable if we consider objects moving at a low speed (this corresponds to a small displacement on the two dimensional image plane) or a relatively high frame rate. Another constraint to be considered is the following: From one frame to the next, a blob may not participate in a splitting and a merging at the same time (parent constraint). This restriction is reasonable assuming a high frame rate where such simultaneous split and merge occurrences are rare [171].

6.12.3 TEMPLATE MATCHING

One remarkable property of Hausdorff-based template matching is its robustness and good performance in locating objects in images of cluttered scenes. The Hausdorff matching score is the Hausdorff distance (HD) between two finite feature point sets. The Hausdorff distance is used to measure the degree of similarity between the extracted contour and a search region in video frames.

The HD computes distance values between two sets of edge points extracted from the object model and a test image. The HD measure is sensitive to degradations such as noise and occlusions, so improved methods have been proposed for image or object matching. The directed distance of the partial HD is defined as:

$$h_K(A,B) = K_{a \in A}^{th} d_B(a) \qquad (32)$$

where $K_{a \in A}^{th}$ denotes the Kth ranked value of $d_B(a)$, and the parameter f is defined by the partial fraction such as $f = \dfrac{K}{N_A}$, with N_A representing the number of points in the set A, whose range is from 0.0 to 1.0. Depending on the fractional value of f, its performance widely varies. Experimentally, when f is about 0.6, good matching results are obtained. This matching Scheme is one of ranked-order statistics methods and yields good results in comparing images that contain outliers and occlusions [172].

Low-level matching algorithms, i.e., algorithms using a distance transform (DT) and a Hausdorff distance (HD) have been investigated because they are simple and insensitive to changes of image characteristics.

The current context, A represents the set of black pixels in the binary image I_T (with white background) and B the set of pixels in a white image that form a black circular template (diameter ≈ 60 pixels for a 640×480 image). Locating the optic disk (OD) amounts to evaluating the Hausdorff distance between the template and the underlying arrangement of pixels in I_T; a perfect match yields a zero distance, which increases as the resemblance weakens. One remarkable property of Hausdorff-based template matching is its robustness and good performance in locating objects in images of cluttered scenes. This capability is certainly valuable for OD localization given the frequent lack of precise disc borders and the presence of vessels coming out of the disc.

In theory, the Hausdorff distance should be computed at each pixel in I_T but in practice, many optimizations help reduce the processing time substantially. One of them is the use of a Voronoi surface (distance transform). Not only does it allow a fast evaluation of the Hausdorff distance at each location, but it also facilitates the implementation of pruning techniques that eliminate areas of the search space around points where the distance is much higher than a given threshold. Pruning is particularly efficient when the input image contains few sparse clusters of black pixels, which is the case when the edge intensity threshold is set appropriately in the previous edge map binarization stage [172–176].

6.12.3.1   Exhaustive Search

Exhaustive search seems to be the only approach at least for small systems. The complete configuration space has been searched up to $N = 32$, the skew-symmetric subspace up to $N = 71$. Fifty days of CPU-time on a special purpose computer have been used for an exhaustive search for binary sequences up to $N = 40$ that minimize max $k\,|C_k|$. Binary sequences $S = \{s_1 = \pm1, ..., s_N\}$ with low off-peak autocorrelations:

$$C_k(S) = \sum_{i=1}^{N-k} s_i s_{i+k} \tag{33}$$

Matching image patches is traditionally done by comparison using sum of squared differences (SSD) or some variant such as normalized cross correlation. It is well understood that the effectiveness of such measures will degrade with significant viewpoint or illumination changes.

One of the most controversial points is whether exhaustive search and template matching are of any help to cope with the desynchronization attack. It is known, in fact, that watermark synchronization through exhaustive search is not only computationally very complex, but it also dramatically increases the false detection probability. At the same time, for the template-matching approach, the probability of a synchronization error must be taken into account [177–180].

The use of cross correlation for template matching is motivated by the distance measure (squared Euclidean distance):

$$d^2_{f,t}(u,v) = \sum_{x,y} \left[f(x,y) - t(x-u, y-v) \right]^2 \tag{34}$$

(the sum is over x, y under the window containing the feature positioned at u, v). In the expansion of d^2:

$$d^2_{f,t}(u,v) = \sum_{x,y}\left[f^2(x,y) - 2f(x,y)t(x-u,y-v) + t^2(x-u,y-v)\right]$$

(35)

the term $\sum t^2(x - u, y - v)$ is constant. If the term $\sum f^2(x, y)$ is approximately constant then the remaining cross correlation term:

$$c(u,v) = \sum_{x,y} f(x,y)t(x-u,y-v)$$ (36)

is a measure of the similarity between the image and the feature [181].

6.12.4 CHAMFER MATCHING

Among the large body of image registration techniques, of particular interest and relevance is chamfer matching (CM), which was originally introduced to match features from two images by means of the minimization of the generalized distance between them. In terms of combining local feature with global matching, the proposed Okapi-Chamfer algorithm is related to the idea of connecting local invariant features with a deformable geometrical model, therefore it can accommodate a wide range of deformations of articulated objects [182, 183].

Although proposed decades ago, chamfer matching remains to be the preferred method when speed and accuracy are considered. There exist several new variants of chamfer matching mainly to improve the cost function using orientation information [184].

Chamfer matching is a popular technique to find the best alignment between two edge maps. Chamfer matching provides a fairly smooth measure of fitness, and can tolerate small rotations, misalignments, occlusions, and deformations. The matching cost can be computed efficiently via a distance transform image $d_{CM}(U, V)$, which specifies the distance from each pixel to the nearest edge pixel.

Let $U = \{u_i\}$ and $V = \{v_j\}$ be the sets of template and query image edge maps respectively. The chamfer distance between U and V is given by the average of distances between each point $u_i \in U$ and its nearest edge in V:

$$d_{CM}(U,V) = \frac{1}{n} \sum_{u_i \in U} \min_{v_i \in V} |u_i - v_j| \qquad (37)$$

where $n = |U|$.

Directional Chamfer Matching (DCM) becomes less reliable in the presence of background clutter. To improve robustness, several variants of chamfer matching have been introduced by incorporating edge orientation information into the matching cost. Each edge point x is augmented with a direction term $\phi(x)$ and the directional chamfer matching (DCM) score is given by

$$d_{DCM}(U,V) = \frac{1}{n} \sum_{u_i \in U} \min_{v_i \in V} |u_i - v_j| + \lambda |\phi(u_i) - \phi(v_j)| \qquad (38)$$

where λ is a weighting factor between location and orientation terms.

To the best of our knowledge, the lowest computational complexity for the existing chamfer matching algorithms is linear in the number of template edge points, even without the directional term [182–184].

6.12.5   KD-TREE SEARCH

KD trees are a generalization of binary search trees. Every node represents a partition of a point set to the two successor nodes. The root represents the whole point cloud and the leaves provide a complete disjunction partition of the points. These leaves are called buckets. Furthermore, every node contains the limits of the represented point set.

A given 3D point needs to be compared with the separating plane in order to decide on which side the search must continue. This procedure is executed until the leaves are reached. There, the algorithm has to evaluate all bucket points. However, the closest point may be in a different bucket, if the distance to the limits is smaller than the one to the closest point in the bucket. Approximate k-d tree search Given an $\varepsilon \rangle 0$, then the point $p \in D$ is the $(1 + \varepsilon)$ approximate nearest neighbor of the point p_q, if

$$\|p - p_q\| \le (1 + \varepsilon)\|p^* - p_q\| \tag{39}$$

where p^* denotes the true nearest neighbor, p has a maximal distance of ε to the true nearest neighbor. Using this notation, in every step the algorithm records the closest point p. The search terminates if the distance to the unanalyzed leaves is larger than

$$\|p - p_q\| \Big/ (1 + \varepsilon) \tag{40}$$

The KD-tree is a well-known space-partitioning data structure for organizing points in k-dimensional space. As an acceleration structure, it has been used in a variety of graphics applications, including triangle culling for ray-triangle intersection tests in ray tracing, nearest photon queries in photon mapping, and nearest neighbor search in point cloud modeling and particle-based fluid simulation (Fig. 6.21). Due to its fundamental importance in graphics, fast kd-tree construction has been a subject of much interest in recent years [186].

One data structure that has been used extensively for nearest neighbors (NN) lookup is the k-d tree. While the "curse of dimensionality" is also a problem for k-d trees, the effects are not as severe. Hash table inefficiency is mainly due to the fact that bin sizes are fixed, whereas

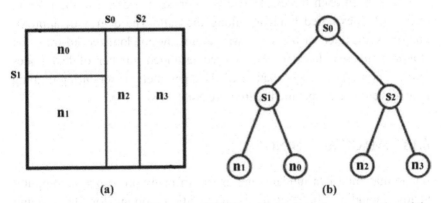

(a)   (b)

FIGURE 6.21   Left: A two-dimensional Kd-Tree. Right: A graph representation of the same Kd-Tree [188].

those in a k-d tree are adaptive to the local density of stored points. Thus, in some cases (high-density region), a hashing approach will have to do a long linear search through many points contained in the bin in which the query point lands; in other cases (low-density), an extensive search through adjacent bins may be required before the best neighbor can be determined. In addition, there is the difficulty of choosing an appropriate bin size [185–188].

6.12.5.1 Best-Bin-First Search

A modification of the k-d tree algorithm called the best-bin-first search method can identify the nearest neighbors with high probability using only a limited amount of computation. Best Bin First (BBF) search, finds the nearest neighbor for a large fraction of queries, and finds a very good neighbor the remaining times. This type of search has wider application than for shape indexing alone; another vision-related use would be for closest point matching in appearance-based recognition [187].

This BBF search strategy provides a dramatic improvement in the NN search for moderate dimensionality, making indexing in these regimes practical. The algorithm initially performs a single traversal through the tree and adds to a priority queue all unexplored branches in each node along the path. Next, it extracts from the priority queue the branch that has the closest center to the query point and it restarts the tree traversal from that branch. In each traversal the algorithm keeps adding to the priority queue the unexplored branches along the path. The degree of approximation is specified in the same way as for the randomized kd-trees, by stopping the search early after a predetermined number of leaf nodes (dataset points) have been examined. This parameter is set during training to achieve the user specified search precision [187–190].

6.13   IMAGE MATCHING

Image matching is a fundamental aspect of many problems in computer vision, including object or image recognition, solving for 3D structure from multiple images, stereo correspondence, and motion tracking [191].

During this stage, the objective is to find all matching images, that is, those that view a common subset of 3D points. Connected sets of image matches will later become 3D models.

From the feature matching step, we have identified images with a large number of matches between them. Since each image could potentially match every other one, this problem appears at first to be quadratic in the number of images. However, we have found it necessary to match each image only to a small number of neighboring images in order to get good solutions for the camera positions. We consider a constant number m images, that have the greatest number of (unconstrained) feature matches with the current image, as potential image matches (we use m = 6).

6.14   RECONSTRUCTION FROM TWO CALIBRATED VIEWS

The interesting feature of the constraint is that although it is nonlinear in the unknown camera poses, it can be solved by two linear steps in closed form. Therefore, in the absence of any noise or uncertainty, given two images taken from calibrated cameras, one can in principle recover camera pose and position of the points in space with a few steps of simple linear algebra.

6.14.1   CAMERA MODEL

A camera is usually described using the pinhole model. The pinhole camera model describes the mathematical relationship between the coordinates of a 3D point and its projection onto the image plane of an ideal pinhole camera, where the camera aperture is described as a point and no lenses are used to focus light. The model does not include, for example, geometric distortions or blurring of unfocused objects caused by lenses and finite sized apertures. It also does not take into account that most practical cameras have only discrete image coordinates. This means that the pinhole camera model can only be used as a first order approximation of the mapping from a 3D scene to a 2D image. Its validity depends on the quality of the camera and, in general, decreases from the center of the image to the edges as lens distortion effects increase.

Some of the effects that the pinhole camera model does not take into account can be compensated for, for example by applying suitable coordinate transformations on the image coordinates, and other effects are sufficiently small to be neglected if a high quality camera is used. This means that the pinhole camera model often can be used as a reasonable description of how a camera depicts a 3D scene, for example in computer vision and computer graphics.

The perspective pinhole camera model has retained the physical meaning of all parameters involved. In particular, the last entry of both x' and X is normalized to 1 so that the other entries may correspond to actual 2D and 3D coordinates (with respect to the metric unit chosen for respective coordinate frames). However, such normalization is not always necessary as long as we know that it is the direction of those homogeneous vectors that matters. For instance, the two vectors $[X,Y,Z,I]^T, [XW,YW,ZW,w]^T \in R^4$ can be used to represent the same point in R^3. Similarly, we can use $[x',y',z']^T$ to represent a point $[x, y, I]^T$ on the 2D image plane as long as $x'/z' = x$ and $y'/z' = y$. However, we may run into trouble if the last entry W or z' happens to be 0.

The following equation expresses the relation between image points and world points. A camera is usually described using the pinhole model. A 3D point $M = [X, Y, Z]^T$ in a Euclidean world coordinate system and the retinal image coordinates $m = [u,v]^T$ are related by the following equation:

$$s\tilde{m} = P\tilde{M} \qquad (41)$$

where s is a scale factor, $\tilde{m} = [u,v,I]^T$ and $\tilde{M} = [X,Y,Z,I]^T$ are the homogeneous coordinates of vector m and M, and P is a 3×4 matrix representing the collineation: $P^3 \rightarrow P^2$. P is called the perspective projection matrix.

Figure 6.10 illustrates this process. The figure shows the case where the projection center is placed at the origin of the world coordinate frame and the retinal plane is at $Z = f = 1$. Then

$$u = \frac{fx}{z}, v = \frac{FY}{Z}$$

$$P = [I_{3\times3} \quad 0_3] \qquad (42)$$

The optical axis passes through the center of projection (camera) C and is orthogonal to the retinal plane. The point c is called the principal point, which is the intersection of the optical axis with the retinal plane. The focal length f of the camera is also shown, which is the distance between the center of projection and the retinal plane (Fig. 6.22).

If only the perspective projection matrix P is available, it is possible to recover the coordinates of the optical center or camera.

The world coordinate system is usually defined as follows: the positive Y-direction is pointing upwards, the positive X-direction is pointing to the right and the positive Z-direction is pointing into the page [192, 193].

The camera calibration matrix is specified by a five parameter upper triangular matrix:

$$K = \begin{pmatrix} f & k & u_0 \\ 0 & rf & v_0 \\ 0 & 0 & 1 \end{pmatrix} \tag{43}$$

An image point x is related to a point in the camera's coordinate system x_c as $x = Kx_c$. Parameter f is the focal length of the camera. The aspect ratio of the camera r depends on the relative scaling of the vertical and

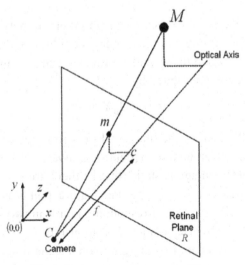

FIGURE 6.22 Perspective projection.

horizontal camera axes. The line from the camera center perpendicular to the image intersects the image at the principal point with coordinates $(u_0 \; v_0)^T$.

The skew, k, is a factor dependent on the physical angle θ between the u and v axes in the sensor array, given by $k = f\cot(\theta)$. Note that radial lens distortion is ignored, but can be corrected in cases where it is significant. In many cases a simplified camera model may be used. A CCD camera, for example, has zero skew ($k = 0$) and a unit aspect ratio ($r = 1$). The resultant simplified or natural camera is

$$K = \begin{pmatrix} f & 0 & u_0 \\ 0 & f & v_0 \\ 0 & 0 & 1 \end{pmatrix} \tag{44}$$

The more general camera model (43) does apply in certain situations however. In many situations the principal point is located near the center of the image, and often can be approximated by the image center. However, it cannot be assumed that this is always the case because photographs (and images) are sometimes cropped before display [88].

6.14.2   EPIPOLAR GEOMETRY

Consider two images of the same image taken from two distinct vantage points. If we assume that the camera is calibrated, the homogeneous image coordinates x and the spatial coordinate X of a point p, with respect to the camera frame, are related by

$$\lambda X = \Pi_0 X \tag{45}$$

where $\Pi_0 = [I, 0]$. That is, the image x differs from the actual 3D coordinates of the point by' an unknown (depth) scale $\lambda \in R^+$. For simplicity, we will assume that the image is static (that is, there are no moving objects) and that the position of corresponding feature points across images is available. If we call x_1, x_2 the corresponding points in two views, they will then be related by a precise geometric relationship.

Assume we have taken two images upon a same object from two different perspectives. Its epipolar geometry is shown in Fig. 6.23.

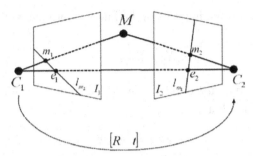

FIGURE 6.23 Epipolar geometry between two views [86].

In the Fig. 6.23, C_1 and C_2 are the centers of two cameras, with m_1 and m_2 as their projection matrices respectively. M is a point in 3D space, m_1 and m_2 are the projections of M in the two image planes respectively. The line connecting C_1 and C_2 is called the baseline, and it intersects the two image planes in points e and e' respectively, called epipoles. The plane π containing the baseline is called the epipolar plane, which intersects the two image planes at two epipolar lines l_1 and l_2, respectively [28, 86, 194–199].

6.14.2.1 The Epipolar Constraint and the Essential Matrix

An orthonormal reference frame is associated with each camera, with its origin 0 at the optical center and the z-axis aligned with the optical axis. The relationship between the 3D coordinates of a point in the inertial "world" coordinate frame and the camera frame can be expressed by a rigid-body transformation. Without loss of generality, we can assume the world frame to be one of the cameras, while the other is positioned and oriented according to a Euclidean transformation $g = (R, T) \in SE(3)$. If we call the 3D coordinates of a point p relative to the two camera frames $X_1 \in R^3$ and $X_2 \in R^3$, they are related by a rigid-body transformation in the following way:

$$X_2 = RX_1 + T \tag{46}$$

Now let $x_1, x_2 \in R^3$ be the homogeneous coordinates of the projection of the same point p in the two image planes. Since $X_i = \lambda_i x_i$, $i = 1, 2$, this equation can be written in terms of the image coordinates X_i and the depths λ_i as

$$\lambda_2 X_2 = R\lambda_1 X_1 + T \tag{47}$$

In order to eliminate the depths λ_i in the preceding equation, premultiply both sides by $\hat{T}$ to obtain

$$\lambda_2 \hat{T} X_2 = \hat{T} R\lambda_1 X_1 \tag{48}$$

Since the vector $\hat{T}x_2 = T \times x_2$ is perpendicular to the vector x_2, the inner product $\langle x_2, \hat{T}x_2 \rangle = x_2{}^T \hat{T}x_2$ is zero. Premultiplying the previous equation by $x_2{}^T$ yields that the quantity $x_2{}^T \hat{T} R\lambda_1 x_1$ is zero. Since $\lambda_1 \rangle 0$, we have proven the following result:

Consider two images x_1, x_2 of the same point p from two camera positions with relative pose $((R, T))$, where $R \in SO(3)$ is the relative orientation and $T \in R^3$ is the relative position. Then x_1, x_2 satisfy

$$\langle x_2, \hat{T} \times Rx_1 \rangle = 0, \text{ or } x_2{}^T \hat{T} Rx_1 = 0 \tag{49}$$

The matrix $E = \hat{T} R \quad R \in R^3$ in the epipolar constraint Eq. (49) is called the *essential matrix*.

In computer vision, the essential matrix is a 3×3 matrix, E, with some additional properties described below, which relates corresponding points in 2D images assuming that the cameras satisfy the pinhole camera model.

Longuet-Higgins' [97] paper includes an algorithm for estimating E from a set of corresponding normalized image coordinates as well as an algorithm for determining the relative position and orientation of the two cameras given that E is known. Finally, it shows how the 3D coordinates of the image points can be determined with the aid of the essential matrix.

Given an essential matrix $E = TR$ that defines an epipolar relation between two images x_1, x_2, we have:

1. The two epipoles e_1, $e_2 \in R^3$, with respect to the first and second camera frames, respectively, are the left and right null spaces of E, respectively:

$$e_2{}^T E = 0, Ee_1 = 0 \tag{50}$$

That is, $e_2 \sim T$ and $e_1 \sim R^T T$. We recall that $\sim$ indicates equality up to a scalar factor.

2. The (co-images of) epipolar lines $l_1, l_2 \in R^3$ associated with the two image points x_1, x_2 can be expressed as

$$l_2 \sim Ex_1, l_1 \sim E^T x_2 \quad \in R^3 \tag{51}$$

where l_1, l_2 are in fact the normal vectors of the epipolar plane expressed with respect to the two camera frames, respectively.

3. In each image, both the image point and the epipole lie on the epipolar line

$$l_i^T e_i = 0, l_i^T x_i = 0, \quad i = 1, 2 \tag{52}$$

6.14.2.2   Estimation of the Essential Matrix

Given a set of corresponding image points it is possible to estimate an essential matrix which satisfies the defining epipolar constraint for all the points in the set. However, if the image points are subject to noise, which is the common case in any practical situation, it is not possible to find an essential matrix which satisfies all constraints exactly.

Depending on how the error related to each constraint is measured, it is possible to determine or estimate an essential matrix which optimally satisfies the constraints of a given set of corresponding image points. The most straightforward approach is to set up a total least squares problem, commonly known as the eight-point algorithm [194–199].

6.14.2.3   Determining R and t from E

Given that the essential matrix has been determined for a stereo camera pair, for example, using the estimation method above, this information can be used for determining also the rotation and translation (up to a scaling) between the two camera's coordinate systems. In these derivations E is seen as a projective element rather than having a well-determined scaling.

The following method for determining R and t is based on performing a *SVD* of E. It is also possible to determine R and t without an *SVD*, for example, following Longuet-Higgins' paper.

An *SVD* of E gives

$$E = U\Sigma V^{T} \tag{53}$$

where U and V are orthogonal 3 × 3 matrices and Σ is a 3 × 3 diagonal matrix with

$$\Sigma = \begin{pmatrix} s & 0 & 0 \\ 0 & s & 0 \\ 0 & 0 & 0 \end{pmatrix} \tag{54}$$

The diagonal entries of Σ are the singular values of E which, according to the internal constraints of the essential matrix, must consist of two identical and one zero value.

To summarize, given E there are two opposite directions which are possible for t and two different rotations which are compatible with this essential matrix. In total, this gives four classes of solutions for the rotation and translation between the two camera coordinate systems. On top of that, there is also an unknown scaling $s)0$ for the chosen translation direction.

It should be noted that the above determination of R and t assumes that E satisfy the internal constraints of the essential matrix. If this is not the case which, for example, typically is the case if E has been estimated from real (and noisy) image data, it has to be assumed that it approximately satisfy the internal constraints. The vector $\hat{t}$ is then chosen as right singular vector of E corresponding to the smallest singular value.

6.14.2.4   3D Points from Corresponding Image Points

The problem to be solved there is how to compute (x_1, x_2, x_3) given corresponding normalized image coordinates (y_1, y_2) and $\left(y_1', y_2'\right)$. If the essential matrix is known and the corresponding rotation and translation transformations have been determined, this algorithm (described in Longuet-Higgins' paper) provides a solution.

Let r_k denote row k of the rotation matrix R:

$$R = \begin{pmatrix} r_1 \\ r_2 \\ r_3 \end{pmatrix} \tag{55}$$

Combining the above relations between 3D coordinates of the two coordinate systems and the mapping between 3D and 2D points described earlier gives

$$y_1' = \frac{x_1'}{x_3'} = \frac{r_1 \cdot (\tilde{x} - t)}{r_3 \cdot (\tilde{x} - t)} = \frac{r_1 \cdot \left(y - \frac{t}{x_3} \right)}{r_3 \cdot \left(\tilde{x} - \frac{t}{x_3} \right)} \tag{56}$$

or

$$x_3 = \frac{\left(r_1 - y_1' r_3 \right) \cdot t}{\left(r_1 - y_1' r_3 \right) \cdot y} \tag{57}$$

Once x_3 is determined, the other two coordinates can be computed as

$$\begin{pmatrix} x_1 \\ x_2 \end{pmatrix} = x_3 \begin{pmatrix} y_1 \\ y_2 \end{pmatrix} \tag{58}$$

The above derivation is not unique. It is also possible to start with an expression for y_2' and derive an expression for x_3 according to

$$x_3 = \frac{\left(r_1 - y_2' r_3 \right) \cdot t}{\left(r_1 - y_2' r_3 \right) \cdot y} \tag{59}$$

In the ideal case, when the camera maps the 3D points according to a perfect pinhole camera and the resulting 2D points can be detected without any noise, the two expressions for x_3 are equal. In practice, however, they are not and it may be advantageous to combine the two estimates of x_3, for example, in terms of some sort of average.

There are also other types of extensions of the above computations which are possible. They started with an expression of the primed image coordinates and derived 3D coordinates in the unprimed system. It is also possible to start with unprimed image coordinates and obtain primed 3D coordinates, which finally can be transformed into unprimed 3D coordinates. Again, in the ideal case the result should be equal to the above expressions, but in practice they may deviate.

A final remark relates to the fact that if the essential matrix is determined from corresponding image coordinate, which often is the case when 3D points are determined in this way, the translation vector t is known only up to an unknown positive scaling. As a consequence, the reconstructed 3D points, too, are undetermined with respect to a positive scaling [28, 86].

6.14.3 PLANAR HOMOGRAPHY

Historically, homographies (and projective spaces) have been introduced to study perspective and projections in Euclidean geometry, and the term "homography," which, etymologically, roughly means "similar drawing" date from this time. At the end of the nineteenth century, formal definitions of projective spaces were introduced, which differed from extending Euclidean or affine spaces by adding points at infinity. The term "projective transformation" originated in these abstract constructions. These constructions divide into two classes that have been shown to be equivalent. A projective space may be constructed as the set of the lines of a vector space over a given field (the above definition is based on this version); this construction facilitates the definition of projective coordinates and allows using the tools of linear algebra for the study of homographies. The alternative approach consists in defining the projective space through a set of axioms, which do not involve explicitly any field (incidence geometry, see also synthetic geometry); in this context, collineations are easier to define than homographies, and homographies are defined as specific collineations, thus called "projective collineations" [28, 86, 200, 201].

A 2D point (x, y) in an image can be represented as a 3D vector $x = (x_1, x_2, x_3)$ where $x = \dfrac{x_1}{x_3}$ and $y = \dfrac{x_2}{x_3}$. This is called the homogeneous representation of a point and it lies on the projective plane P^2. A homography is an invertible mapping of points and lines on the projective plane P^2.

Other terms of this transformation include collineation, projectivity, and planar projective transformation. Hartley and Zisserman [11] provide the specific definition that a homography is an invertible mapping from P^2 to itself such that three points lie on the same line if and only if their mapped points are also collinear. They also give an algebraic definition by proving the following theorem: A mapping from $p^2 \rightarrow P^2$ is a projectivity if and only if there exists a nonsingular 3×3 matrix H such that for any point in P^2 represented by vector x it is true that its mapped point equals Hx. This tells us that in order to calculate the homography that maps each x_i to its corresponding x_1' it is sufficient to calculate the 3×3 homography matrix H. It should be noted that H can be changed by multiplying by an arbitrary nonzero constant without altering the projective transformation. Thus H is considered a homogeneous matrix and only has 8 degrees of freedom, even though it contains 9 elements. This means there are 8 unknowns that need to be solved for. Typically, homographies are estimated between images by finding feature correspondences in those images. The most commonly used algorithms make use of point feature correspondences, though other features can be used as well, such as lines or conics.

$$\underbrace{\begin{bmatrix} wx' \\ wy' \\ w \end{bmatrix}}_{x'} = \underbrace{\begin{bmatrix} * & * & * \\ * & * & * \\ * & * & * \end{bmatrix}}_{H} \underbrace{\begin{bmatrix} x \\ y \\ 1 \end{bmatrix}}_{x} \tag{60}$$

$$\lambda_2 x_2' = H \lambda_1 x_1' \quad \lambda_2 x_2' = HX \quad X = \begin{bmatrix} X & Y & 1 \end{bmatrix}^T \tag{61}$$

$$\widehat{x_2'} H_1 x_1' = 0 \tag{62}$$

The homography transformation has 8 degrees of freedom and there are other simpler transformations that still use the 3×3 matrix but contain specific constraints to reduce the number of degrees of freedom (Fig. 6.24) [200].

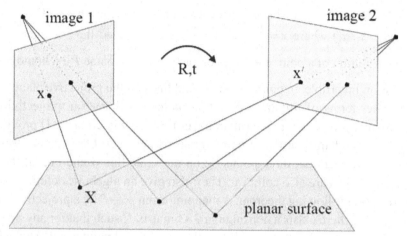

FIGURE 6.24   The geometry of a homography mapping [201].

6.14.3.1   Relationships Between the Homography and the Essential Matrix

In practice, especially when the image is piecewise planar, we often need to compute the essential matrix E with a given homography H computed from some four points known to be planar; or in the opposite situation, the essential matrix E may have been already estimated using points in general position, and we then want to compute the homography for a particular (usually smaller) set of coplanar points. We hence need to understand the relationship between the essential matrix E and the homography H [200–202].

6.14.4   FUNDAMENTAL MATRIX

Structure from uncalibrated images only leads to a projective reconstruction. This image data is usually in the form of corners (high curvature points), as they can be easily represented and manipulated in projective geometry [28].

The essential matrix can be seen as a precursor to the fundamental matrix. Both matrices can be used for establishing constraints between matching image points, but the essential matrix can only be used in relation to calibrate cameras since the inner camera parameters must be known in order to achieve the normalization. If, however, the cameras are calibrated the essential matrix can be useful for determining both the relative position and orientation between the cameras and the 3D position of corresponding image points (Fig. 6.25).

6.14.4.1 Fundamental Matrix Estimation

Let be F the fundamental matrix. Some important equations are listed here:

$$x_2'^{T} F x_1' = 0 \tag{63}$$

$$\begin{aligned} l_1 &\sim F^T x_2' & l_i^T x_i' &= 0 & l_2 &\sim F^T x_1' \\ F e_1 &= 0 & l_i^T e_i &= 0 & e_2^T F &= 0 \end{aligned} \tag{64}$$

In the case of calibrated cameras, becomes the essential matrix, which relates the normalized points in two images.

After finding a set of matching points from two images using key point descriptors, we find the fundamental matrix using normalized 8-point algorithm [28, 86].

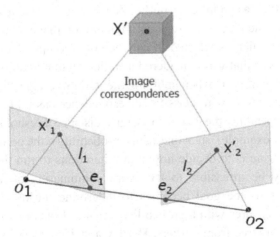

FIGURE 6.25 Project of 2D point to 3D point with uncalibrated camera.

6.15   ROBUST ESTIMATION

Robust methods for determining the fundamental matrix are especially important when dealing with real image data, which is able to detect outliers in the correspondence. As the fundamental matrix has only seven degrees of freedom, it is possible to estimate F directly using only 7 point matches. In general, more than 7 point matches are available and a method for solving the fundamental matrix using 8 point matches. The points in both images are usually subject to noise and therefore a minimization technique is implemented. A robust method is very useful as the technique will ignore these false matches in the estimation of the fundamental matrix. The two most commonly used approaches of robust estimation are RANSAC and LMS [86, 200].

6.15.1   RANSAC

Random Sample Consensus (RANSAC) is the most commonly used robust estimation method for homographies and this method makes decisions based on the number of data points within a distance threshold [203]. The idea of the algorithm is pretty simple; for a number of iterations, a random sample of four correspondences is selected and a homography H is computed from those four correspondences. Each other correspondence is then classified as an inlier or outlier depending on its concurrence with H. After all of the iterations are done, the iteration that contained the largest number of inliers is selected. H can then be recomputed from all of the correspondences that was considered as inliers in that iteration.

One important issue when applying the RANSAC algorithm described above is to decide how to classify correspondences as inliers or outliers. Statistically speaking, the goal is to assign a distance threshold, t, (between x' and Hx, for example), such that with a probability α the point is an inlier.

Another issue is to decide how many iterations to run the algorithm. It will likely be infeasible to try every combination of 4 correspondences, and thus the goal becomes to determine the number of iterations, N, that ensures with a probability p that at least one of the random samples will be free from outliers. Hartley and Zisserman [54] show that

$$N = \frac{\log(1-p)}{\log\left(1-(1-\in)^s\right)},$$ where $\in$ is the probability that a sample correspondence is an outlier and s is the number of correspondences used in each iteration, 4 in this case. If $\in$ is unknown, the data can be probed to adaptively determine $\in$ and N.

6.15.2 LEAST MEDIAN OF SQUARES REGRESSION

This is one way to deal with the fact that the sum of squared difference algorithms such as the Algebraic distance version of the DLT is not very robust with respect to outliers. There is a lot of research on the topic of ways to improve the robustness of regression methods. One example would be to replace the squared distance with the absolute value of the distance. This improves the robustness since outliers aren't penalized as severely as when they are squared. A popular approach with respect to homography estimation is Least Median of Squares (LMS) estimation. As described in Ref. [204], this method replaces the sum with the median of the squared residuals. LMS works very well if there are less than 50% outliers and has the advantage over RANSAC that it requires no setting of thresholds or a priori knowledge of how much error to expect (unlike the setting of the t and $\in$ parameters in RANSAC).

The major disadvantage of LMS is that it would be unable to cope with more than half the data being outliers. In this case, the median distance would be to an outlier correspondence [200].

6.15.3 FUTURE WORK

It is interesting to note that RANSAC itself is not a great solution when there is a high percentage of outliers, as its computational cost can blow up with the need for too many iterations. While RANSAC and LMS are the most commonly used methods for robust homography estimation, there may be an opening for research into whether there are other methods aside from those two that would do well in the presence of a very high number of outlier data. One potentially successful approach could be to try to fit a lower order transformation from a set of correspondences, such as a similarity transform. While this would be unlikely to perfectly segment outliers from inliers, it could be useful for removing the obvious outliers. By disproportionally removing more outliers than inliers, a situation where RANSAC would have previously failed can be brought into a realm where it could work [200].

Also M. Brown et al. [165] parameterize each camera using 7 parameters. These are a rotation vector $\Theta_i = \begin{bmatrix} \Theta_{i1} & \Theta_{i2} & \Theta_{i3} \end{bmatrix}$ translation $t_i = \begin{bmatrix} t_{i1} & t_{i2} & t_{i3} \end{bmatrix}$ and focal length f_i. The calibration matrix is then

$$K_i = \begin{bmatrix} f_i & 0 & 0 \\ 0 & f_i & 0 \\ 0 & 0 & 1 \end{bmatrix} \tag{65}$$

and the rotation matrix (using exponential representation)

$$R_i = e^{[\theta_i]_x}, \quad [\theta_i]_x = \begin{bmatrix} 0 & -\theta_{i3} & \theta_{i2} \\ \theta_{i3} & 0 & -\theta_{i1} \\ -\theta_{i2} & \theta_{i1} & 0 \end{bmatrix} \tag{66}$$

Each pairwise image match adds four constraints on the camera parameters while adding three unknown structure parameters $X = [X_1\ X_2\ X_3]$

$$\tilde{u}_i = K_i X_{ci} \tag{67}$$

$$\tilde{u}_j = K_i X_{cj} \tag{68}$$

$$X_{ci} = R_i X + t_i \tag{69}$$

$$X_{cj} = R_j X + t_j \tag{70}$$

where $\tilde{u}_i, \tilde{u}_j$ are the homogeneous image positions in camera i and j respectively. The single remaining constraint (4 equations minus 3 unknowns = 1 constraint) expresses the fact that the two camera rays $\tilde{p}_i, \tilde{p}_j$ and the translation vector between camera centers t_{ij} are coplanar, and hence their scalar triple product is equal to zero

$$\tilde{p}_i^T \left[t_{ij} \right] \times \tilde{p}_j = 0 \qquad (71)$$

Writing $\tilde{p}_i, \tilde{p}_j$ and t_{ij} in terms of camera parameters

$$\tilde{p}_i = R_i^T K_i^{-1} \tilde{u}_i \qquad (72)$$

$$\tilde{p}_j = R_j^T K_j^{-1} \tilde{u}_j \qquad (73)$$

$$t_{ij} = R_j^T t_j - R_i^T t_i \qquad (74)$$

And substituting in Eq. (75) gives

$$\tilde{u}_i^T F_{ij} \tilde{u}_j = 0 \qquad (75)$$

Where

$$F_{ij} = K_i^{-T} R_i \left[R_j^T t_j - R_i^T t_i \right] \times R_j^T K_j^{-1} \qquad (76)$$

This is the well-known epipolar constraint. Image matching entails robust estimation of the fundamental matrix F_{ij}. Since Eq. (75) is non-linear in the camera parameters, it is commonplace to relax the nonlinear constraints and estimate a general 3×3 matrix F_{ij}. This enables a closed form solution via SVD.

In this work are used RANSAC to robustly estimate F and hence find a set of inliers that have consistent epipolar geometry. An image match is declared if the number of RANSAC inliers $n_{inliers} \rangle n_{match}$, where the minimum number of matches n_{match} is a constant (typically around 20). Then 3D objects/images are identified as connected components of the image matches [165].

6.16   BUNDLE ADJUSTMENT

Bundle adjustment is the problem of refining a visual reconstruction to produce jointly optimal 3D structure and viewing parameter (camera pose and/or calibration) estimates. Optimal means that the parameter estimates

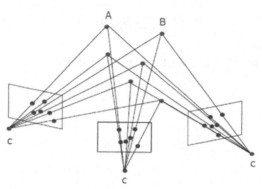

FIGURE 6.26   Jointly optimal 3D structure bundle adjustment.

are found by minimizing some cost function that quantifies the model fitting error, and jointly that the solution is simultaneously optimal with respect to both structure and camera variations. The name refers to the 'bundles' of light rays leaving each 3D feature and converging on each camera center, which are 'adjusted' optimally with respect to both feature and camera positions. Equivalently unlike independent model methods, which merge partial reconstructions without updating their internal structure all of the structure and camera parameters are adjusted together 'in one bundle.' Bundle adjustment is really just a large sparse geometric parameter estimation problem, the parameters being the combined 3D feature coordinates, camera poses and calibrations. Almost everything that we will say can be applied to many similar estimation problems in vision, photogrammetry, industrial metrology, surveying and geodesy (Fig. 6.26) [165, 205–207].

6.17   VISUALIZATION

Visualization of the model during image acquisition allows the operator to interactively verify that an adequate set of input images has been collected for the modeling task, while automatic image selection keeps storage requirements to a minimum [86, 208].

Single 3D points cannot provide a global illustration about the structure of the object. Thereby, the creation of the 3D object model is a requirement in order to depict the formation and the real conditions of the object.

Furthermore, the 3D reconstruction is a requirement for further processes such as the application of visualization and graphics techniques. In any case the 3D object model can be developed under standard or special CAD software [209].

There are many different visualization methods from simple hand drawings, to CAD designs, GIS systems, 3D representations, animations and walk-throughs or even stereoscopic representations in virtual reality applications.

The 3D representation has to provide adequate geometry characteristics and detailed enough textures for the archaeologists to be able to work on them. While the process of creating the virtual models can be complex, there are various techniques that try to automate the whole process as much as possible. During the visualization process, there are two options, to visualize the actual objects or to visualize the reconstructed objects [209].

When all pairs of images have been matched, we can construct an image connectivity graph to represent the connections between the images in the collection. An image connectivity graph contains a node for each image, and an edge between any pair of images that have matching features. To create this visualization, the graph was embedded in the plane using the Neato tool in the Graphviz graph visualization toolkit. Neato works by modeling the graph as a mass-spring system and solving for an embedding whose energy is a local minimum. The image connectivity graph for this collection has several notable properties. There is a large, dense cluster in the center of the graph that consists of photos that are mostly wide-angle, frontal, well-lit shots of the fountain. Other images, including the Bleaf [nodes] corresponding for tightly cropped detail, and nighttime images, are more loosely connected to this core set [13, 210, 211].

6.18   EPIPOLAR RECTIFICATION

Image rectification is an important component of stereo computer vision algorithms. Rectification is a process used to facilitate the analysis of a stereo pair of images by making it simple to enforce the two view geometric constraints. We assume that a pair of 2D images of a 3D object or environment is taken from two distinct viewpoints and their epipolar geometry has been determined. Corresponding points between the two images must satisfy the so-called epipolar constraint. For a given point in one image, we have to search for its correspondence in the other image along an epipolar line. In general, epipolar lines are not aligned with coordinate axis and are not parallel. Such searches are time consuming since we must compare pixels on skew lines in image space. These types of algorithms can be simplified and made more efficient if epipolar lines are axis aligned and parallel. This can be realized by applying 2D projective transforms, or homographies, to each image. This process is known as image rectification. Generally, the rectification process expects to be provided with two rectangular images as well as a fundamental matrix and a set of point matches such as those used to calculate the fundamental matrix and also rectification can be used to recover 3D structure from an image pair without appealing to 3D geometric notions like cameras (Fig. 6.27) [195, 197].

Some previous techniques for finding image rectification homographies involve 3D constructions. These methods find the 3D line of intersection between image planes and project the two images onto the plane containing this line that is parallel to the line joining the optical centers. While this approach is easily stated as a 3D geometric construction, its realization in practice is somewhat more involved and no consideration is given to other more optimal choices. A strictly 2D approach that does attempt to optimize the distorting effects of image rectification can be found in. Their distortion minimization criterion is based on a simple geometric heuristic which may not lead to optimal solutions. The important advantage of rectification is that computing correspondences are made simpler, because search is done along the horizontal lines of the rectified images [88, 195–199].

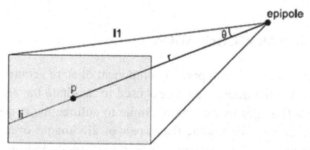

FIGURE 6.27   Epipolar rectification. A point p is encoded by a pair of (r; θ) [203].

6.19   DENSE MATCHING

A dense matching algorithm is integrated as preprocessing. It computes the corresponding information for only those sufficiently textured areas. Matching is propagated from the most reliably matched pixels to their neighbors. Propagation is stopped when the texture cue is not sufficient [212–214].

Traditional dense matching problems such as stereo or optical flow deal with the "instance matching" scenario, in which the two input images contain different viewpoints of the same image or object. More recently, researchers have pushed the boundaries of dense matching to estimate correspondences between images with different images or objects. This advance beyond instance matching leads to many interesting new applications, such as semantic image segmentation, image completion, image classification, and video depth estimation [191].

There are two major challenges when matching generic images: image variation and computational cost. Compared to instances, different images and objects undergo much more severe variation in appearance, shape, and background clutter. These variations can easily confuse low level matching functions. At the same time, the search space is much larger, since generic image matching permits no clean geometric constraints. Without any prior knowledge of the images' spatial layout, in principle, we must search every pixel to find the correct match [215–217].

Recent innovations in matching algorithms considerably improved the quality of elevation data, generated automatically from aerial images. Traditional matching, originally introduced more than two decades ago, usually applies feature based algorithms. These algorithms first extract feature points and then search the corresponding features in the overlapping images. The restriction to matches of selected points usually provides correspondences at high certainty. However, feature based matching was also introduced to avoid problems due to limited computational resources. In contrast, recent algorithms aim on dense, pixel-wise matches. By these means 3D point clouds and digital surface models (DSM) are generated at a resolution, which corresponds to the ground sampling distance (GSD) of the original images. To compute pixel matches even for regions with very limited texture, additional constraints are required [211, 218].

6.20 TRIANGULATION

Triangulation is the process of determining the location of a point by measuring angles to it from known points at either end of a fixed baseline, rather than measuring distances to the point directly (trilateration). The point can then be fixed as the third point of a triangle with one known side and two known angles. The 3D position of the points can be obtained easily by triangulation. The process of triangulation is needed to find the intersection of two known rays in space. Due to measurement noise in images and some inaccuracies in the calibration matrices, these two rays will not generally meet in a unique point [28, 219–224].

The triangulation problem is a small cog in the machinery of computer vision, but in many applications of scene reconstruction it is a critical one, on which ultimate accuracy depends [76].

By the investigation of Voronoi diagrams goes the investigation of related constructs. Among them, the Delaunay triangulation is most prominent. It contains a (straight-line) edge connecting two sites in the plane if and only if their Voronoi regions share a common edge. The structure was introduced by Voronoi [1908] for sites that form a lattice and was extended by Delaunay to irregularly placed sites by means of the empty-circle method: Consider all triangles formed by the sites such that the circumcircle of each

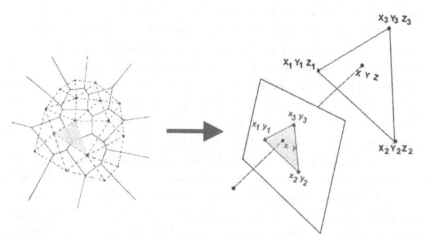

FIGURE 6.28   Corresponding triangles in 2D image point and 3D space point (for affine coordinates).

triangle is empty of other sites. The set of edges of these triangles gives the Delaunay triangulation of the sites (Fig. 6.28) [207, 224].

6.21   TEXTURE MAPPING

Texture mapping has traditionally been used to add realism to computer graphics images. In its basic form, texture mapping lays an image (the texture) onto an object in a image. In recent years, this technique has moved from the domain of software rendering systems to that of high performance graphics hardware, also Texture mapping ensures that "all the right things" happen as a textured polygon is transformed and rendered.

In order to render the image from novel viewpoints, we need a model of its, surfaces, so that we can texture-map images onto it. Clearly, the handful of point features we have reconstructed so far is not sufficient [86].

More general forms of texture mapping generalize the image to other information; an "image" of altitudes, for instance, can be used to control shading across a surface to achieve such effects as bump-mapping [225, 226].

The texture mapping can be done as follows:

1. back project the 3D mesh, a set of wire-frame patches, into each frame.

2. extract texture patches (photometric information) of each wire-frame patches

3. use photometric and geometric information, i.e. the angles between the line of sight of views and the normal of the wire-frame patch), to create the mapping texture patch.

4. map the texture patches to the corresponding wire-frame patches.

One simple use of texture mapping is to draw antialiased points of any width. In this case the texture image is of a filled circle with a smooth (antialiased) boundary. When a point is specified, its coordinates indicate the center of a square whose width is determined by the point size. The texture coordinates of the square's corners are those corresponding to the corners of the texture image. This method has the advantage that any point shape may be accommodated simply by varying the texture image [203, 226].

6.22   BETTER METHOD FOR 3D RECONSTRUCTION OF NANOPOROUS MEMBRANE

Several instrumental characterization techniques have been suggested to obtain 3D volume images of pore space, such as X-ray computed micro tomography and magnetic resonance computed micro tomography. However, these techniques may be limited by their resolution. So, the 3D stochastic reconstruction of porous media from statistical information (produced by analysis of 2D photomicrographs) has been suggested. Although pore network models can be two- or three-dimensional, 2D image analysis, due to their restricted information about the whole microstructure, was unable to predict morphological characteristics of porous membrane. Therefore 3D reconstruction of porous structure will lead to significant improvement in predicting the pore characteristics. Recently research work has focused on the 3D image analysis of porous membranes [227–232].

6.23   CONCLUSIONS

In recent years, great efforts have been devoted to nanoporous membranes. As a conclusion, much progress has been made in the preparation and characterization of porous media. Among several porous membranes, electrospun nanofibrous membranes, due to the high porosity, large surface area-to-volume ratios, small pores, and fine fiber diameter, have gained increasing attention. Useful techniques for evaluation of the pore characteristics of porous membranes are reviewed. Image analysis techniques have been suggested as a useful method for characterization of porous media due to its convenience in detecting individual pores. Although image analysis of SEM micrographs for geometrical characterization is useful for measuring the total porosity, pore shape, pore size and pore size distribution of relatively thin nonwovens, it cannot be applied to multilayer electrospun fibrous analysis. Another problem encounter to this method is that it is not possible to measure 3D pore characteristics of the membrane and it is limited on relatively small fields of view. It is believed that the 3D reconstruction of porous media, from the information obtained from a 2D analysis of photomicrographs, will bring a promising future to nanoporous membranes.

The topic of obtaining 3D models from images is a new research field in pore analysis of nanofibrous membrane. A set of techniques for creating a 3D representation of a view from one or more 2D images can be used to image-based modeling. A solution to this lack of 3D content is to convert existing 2D material to 3D. A detailed review on 3D image reconstruction from two views of single 2D image has been done in this contribution. The chapter concisely demonstrated that there are three steps for 3D reconstruction, comprising estimation of the epipolar geometry existing between the 2D image pair, estimation of the affine geometry, and also camera calibration. The obtained results for both the camera calibration and reconstruction showed that it is possible to obtain a 3D model directly from features in the images.

KEYWORDS

- **2D image**
- **3D image analysis**
- **3D reconstruction**
- **nanofibrous membrane**
- **pore structure**

REFERENCES

1. Barhate, R. S., Ramakrishna, S., "Nanofibrous filtering media: filtration problems and solutions from tiny materials," Journal of membrane science, Vol. 296, No. 1, 1–8, 2007.
2. Cho, D., Naydich, A., Frey, M. W., Joo, Y. L., "Further improvement of air filtration efficiency of cellulose filters coated with nanofibers via inclusion of electrostatically active nanoparticles," Polymer, Vol. 54, No. 9, 2364–2372, 2013.
3. Jena, A., Gupta, K., "Characterization of pore structure of filtration media" Fluid/ Particle Separation Journal, Vol. 14, No. 3, 227–241, 2002.
4. Bagherzadeh, R., Latifi, M., Najar, S. S., Kong, L., "Three dimensional pore structure analysis of Nano/Microfibrous scaffolds using confocal laser scanning microscopy," Journal of Biomedical Materials Research Part A, Vol. 101, No. 3, 765–774, 2013.

5. Ghasemi Mobarakeh, L., Semnani, D., Morshed, M., "A Novel Method for Porosity Measurement of Various Surface Layers of Nanofibers Mat Using Image Analysis for Tissue Engineering Applications," Journal of Applied Polymer Science, Vol. 106, 2536–2542, 2007.

6. Sreedhara, S. S., Rao Tata, N., "A Novel Method for Measurement of Porosity in Nanofiber Mat using Pycnometer in Filtration," Journal of Engineered Fibers and Fabrics, Vol. 8, Issue 4, 2013.

7. Strange, J. H., Rahman, M., Smith, E. G., "Characterization of porous solids by NMR," Physical review letters, Vol. 71, No. 21, 3589, 1993.

8. Ziabari, M., Mottaghitalab, V., Haghi, A. K., "Evaluation of electrospun nanofiber pore structure parameters" Korean Journal of Chemical Engineering, Vol. 25, No. 4, 923–932, 2008.

9. Nimmo, J. R., "Porosity and pore size distribution" Encyclopedia of Soils in the Environment, Vol. 3, 295–303, 2004.

10. He, M., Zeng, Y., Sun, X., Harrison, D. J., "Confinement effects on the morphology of photopatterned porous polymer monoliths for capillary and microchip electrophoresis of proteins" Electrophoresis, Vol. 29, No. 14, 2980–2986, 2008.

11. Al-Kharusi, A. S., Martin J. B., "Network extraction from sandstone and carbonate pore space images" Journal of Petroleum Science and Engineering, Vol. 56, No. 4, 219–231, 2007.

12. Al-Raoush, R. I., Willson, C. S., "Extraction of physically realistic pore network properties from three-dimensional synchrotron X-ray microtomography images of unconsolidated porous media systems" Journal of hydrology, Vol. 300.No. 1, 44–64, 2005.

13. Al-Raoush, R., "Change in microstructure parameters of porous media over representative elementary volume for porosity" Particulate Science and Technology, Vol. 30, No. 1, 1–16, 2012.

14. Miao, J., Ishikawa, T., Johnson, B., Anderson, E. H., Lai, B., Hodgson, K. O., "High resolution 3D x-ray diffraction microscopy," Physical review letters, Vol. 89, No. 8, 088303, 2002.

15. Rollett, A. D., Lee, S. B., Campman, R., Rohrer, G. S. "Three-dimensional characterization of microstructure by electron back-scatter diffraction," Annu. Rev. Mater. Res., Vol. 37, 627–658, 2007.

16. Prior, D. J., Boyle, A. P., Brenker, F., Cheadle, M. C., Day, A., Lopez, G., Zetterström, L, "The application of electron backscatter diffraction and orientation contrast imaging in the SEM to textural problems in rocks," American Mineralogist, Vol. 84, 1741–1759, 1999.

17. Davies, P. A., Randle, V., "Combined application of electron backscatter diffraction and stereophotogrammetry in fractography studies," Journal of microscopy, Vol. 204, No. 1, 29–38, 2001.

18. Uchic, M. D., Groeber, M. A., Dimiduk, D. M., Simmons, J. P., "3D microstructural characterization of nickel superalloys via serial-sectioning using a dual beam FIB-SEM," Scripta Materialia, Vol. 55, No. 1, 23–28, 2006.

19. Khokhlov, A. G., Valiullin, R. R., Stepovich, M. A., Kärger, J., "Characterization of pore size distribution in porous silicon by NMR cryoporosimetry and adsorption methods," Colloid Journal, Vol. 70, No. 4, 507–514, 2008.

20. Remondino, F., El-Hakim, S. "Image based 3D Modeling: A Review," The Photogrammetric Record, Vol. 21, No. 115, 269–291, 2006.
21. Kushal, A. M., Bansal, V., Banerjee, S., "A simple method for interactive 3D reconstruction and camera calibration from a single view," ICVGIP, 2002.
22. Sturm, P.F., Maybank, S. J., "A Method for Interactive 3D Reconstruction of Piecewise Planar Objects from Single Images," The 10th British Machine Vision Conference (BMVC '99) 265–274, 1999.
23. Criminisi, A., Reid, I., Zisserman, A., "Single View Metrology," International Journal of Computer Vision, Vol. 40, No. 2, 123–148, 2000.
24. Debevec, P. E., "Modeling and Rendering Architecture from Photographs," PhD thesis, University Of California At Berkeley, 1996.
25. Grossmann, E., Ortin, D., Santos-Victor, J., "Single and Multi-View Reconstruction of Structured Images," the 5th Asian Conference on Computer Vision, 23–25, 2002.
26. Liebowitz, D., Criminisi, A., Zisserman, A., "Creating Architectural Models from Images," the Eurographics Association and Blackwell Publishers, Vol. 18, No.3, 1999.
27. Wilczkowiak, M., Boyer, E., Sturm, P., "Camera Calibration and 3D Reconstruction from Single Images Using Parallelepipeds," 8th International Conference on Computer Vision (ICCV '01), Vol. 1, 142–148, 2001.
28. Henrichsen, A., "3D Reconstruction and Camera Calibration from 2D Images," MSc Thesis, University Of Cape Town, 2000.
29. Hartley, R., Silpa-Anan, C., "Reconstruction from two views using approximate calibration," Proc. 5th Asian Conf. Comput. Vision, 2002.
30. Snavely, N., Simon, I., Goesele, M., Szeliski, R., Seitz, S. M., "Image Reconstruction and Visualization from Community Photo Collections," Proceedings of the IEEE, Vol. 98, No. 8, 2010.
31. Zhang, L., Member, S., Vázquez, C., Knorr, S., "3D-TV Content Creation: Automatic 2D-to-3D Video Conversion," IEEE Transactions on Broadcasting, Vol. 57, No. 2, 2011.
32. Andal, F. A., Taubin, G., Goldenstein, S., "Vanishing Point Detection by Segment Clustering on the Projective Space," Trends and Topics in Computer Vision, Lecture Notes in Computer Science Volume 6554, 324–33, 2012.
33. Ourselin, S., Roche, A., Subsol, G., Pennec, X., Ayache, N., "Reconstructing a 3D structure from serial histological sections," Image and Vision Computing, Vol. 19, 25–31, 2000.
34. Barhate, R. S., Loong, C. K., Ramakrishna, S., "Preparation and characterization of nanofibrous filtering media," Journal of Membrane Science, Vol. 283, No. 1, 209–218, 2006.
35. Barhate, R. S., Ramakrishna, S., "Nanofibrous filtering media: filtration problems and solutions from tiny materials," Journal of Membrane Science, Vol. 296, No. 1, 1–8, 2007.
36. Zong, X., Kim, K., Fang, D., Ran, S., Hsiao, B. S., Chu, B., "Structure and process relationship of electrospun bioabsorbable nanofiber membranes," Polymer, Vol.43, No. 16, 4403–4412, 2002.
37. Burger, C., Hsiao, B. S., Chu, B., "Nanofibrous materials and their applications." Annu. Rev. Mater. Res. Vol.36, 333–368, 2006.

38. Gibson, P., Schreuder-Gibson, H., Rivin, D., "Transport properties of porous membranes based on electrospun nanofibers" Colloids and Surfaces A: Physicochemical and Engineering Aspects, Vol. 187, 469–481, 2001.
39. Pham, Q. P., Sharma, U., Mikos, A. G., "Electrospinning of polymeric nanofibers for tissue engineering applications: a review" Tissue engineering, Vol. 12, No. 5, 1197–1211, 2006.
40. Gopal, R., Kaur, S., Ma, Z., Chan, C., Ramakrishna, S., Matsuura, T., "Electrospun nanofibrous filtration membrane," Journal of Membrane Science, Vol. 281, No. 1, 581–586, 2006.
41. Feng, C., Khulbe, K. C., Matsuura, T., Gopal, R., Kaur, S., Ramakrishna, S., Khayet, M., "Production of drinking water from saline water by air-gap membrane distilation using polyvinylidene fluoride nanofiber membrane" Journal of Membrane Science, Vol. 311, No. 1, 1–6, 2008.
42. Zander, N. E., "Hierarchically Structured Electrospun Fibers," Polymers, Vol. 5, No. 1, 19–44, 2013.
43. Bhardwaj, N., Kundu, S. C., "Electrospinning: a fascinating fiber fabrication technique" Biotechnology advances, Vol. 28, No. 3, 325–347, 2010.
44. Wang, N., Burugapalli, K., Song, W., Halls, J., Moussy, F., Ray, A., Zheng, Y., "Electrospun fibro-porous polyurethane coatings for implantable glucose biosensors," Biomaterials, Vol. 34, No. 4, 888–901, 2013.
45. Jung, H. R., Ju, D. H., Lee, W. J., Zhang, X., Kotek, R., "Electrospun hydrophilic fumed silica/polyacrylonitrile nanofiber-based composite electrolyte membranes," Electrochimica Acta, Vol. 54, No. 13, 3630–3637, 2009.
46. Gong, Z., Ji, G., Zheng, M., Chang, X., Dai, W., Pan, L., Zheng, Y., "Structural characterization of mesoporous silica nanofibers synthesized within porous alumina membranes," Nanoscale research letters, Vol. 4, No. 11, 1257–1262, 2009.
47. Wang, Y., Zheng, M., Lu, H., Feng, S., Ji, G., Cao, J. "Template synthesis of carbon nanofibers containing linear mesocage arrays," Nanoscale research letters, Vol. 5, No. 6, 913–916, 2010.
48. Yin, G., "Analysis of Electrospun Nylon 6 Nanofibrous Membrane as Filters," Journal of Fiber Bioengineering and Informatics, Vol. 3, No.3, 137–141, 2010.
49. Lee, J. B., Jeong, S. I., Bae, M. S., Yang, D. H., Heo, D. N., Kim, C. H., Kwon, I. K., "Highly porous electrospun nanofibers enhanced by ultrasonication for improved cellular infiltration," Tissue Engineering Part A, Vol. 17, No. 21–22, 2695–2702, 2011.
50. Kim, G., Kim, W., "Highly porous 3D nanofiber scaffold using an electrospinning technique." Journal of Biomedical Materials Research Part B: Applied Biomaterials, Vol. 81, No. 1, 104–110, 2007.
51. Gregg, A. "The application of nanofibrous membranes with antimicrobial agents as filters," MSc thesis, Kansas State University, 2010.
52. Nasreen, S. A. A. N., Sundarrajan, S., Nizar, S. A. S., Balamurugan, R., Ramakrishna, S., "Advancement in Electrospun Nanofibrous Membranes Modification and Their Application in Water Treatment," Membranes, Vol. 3, No. 4, 266–284, 2013.
53. Grafe, T., Graham, K., "nanofibers web from electrospinning," Nonwovens in Filtration – Fifth International Conference, Stuttgart, Germany, 2003.
54. Graham, K., Ouyang, M., Raether, T., Grafe, T., McDonald, B., Knauf, P., "Polymeric nanofibers in air filtration applications," In Fifteenth Annual Technical Conference & Expo of the American Filtration and Separations Society, Galveston, Texas, 9–12, 2002.

55. Yoon, K., Kim, K., Wang, X., Fang, D., Hsiao, B. S., Chu, B., "High flux ultrafiltration membranes based on electrospun nanofibrous PAN scaffolds and chitosann coating," Polymer, Vol. 47, No. 7, 2434–2441, 2006.

56. Fatarella, E., Iversen, V., Grinwis, S., Paulussen, S., "Characterization Test On Selected NonWoven Web," Project cofunded by the European Commission within the Sixth Framework Program, 2009.

57. Desai, K., Kit, K., Li, J., Davidson, P. M., Zivanovic, S., Meyer, H., "Nanofibrous chitosan nonwovens for filtration applications," Polymer, Vol. 50, No. 15, 3661–3669, 2009.

58. Matrecano, M., "Porous Media Characterization by Micro-Tomographic Image Processing," PhD thesis, Università Degli Studi Di Napoli "Federico Ii," 2011.

59. Glover, P., "Formation Evaluation MSc Course Notes," Aberdeen University, 84–94, 2001.

60. Patel, S. U., Manzo, G. M., Patel, S. U., Kulkarni, P. S., Chase, G. G., "Permeability of electrospun superhydrophobic nanofiber mats," Journal of Nanotechnology, 2012.

61. Saar, M. O., Manga, M., "Permeability-porosity relationship in vesicular basalts," Geophysical Research Letters, Vol. 26, No. 1, 111–114, 1999.

62. Saar, M. O., "The Relationship between Permeability, Porosity and Micro Structure in Vesicular Basalts," MSc Thesis, University of Oregon, 1998.

63. Nelson, P. H., "Permeability-porosity relationships in sedimentary rocks," the log analyst, Vol. 35, No. 3, 1994.

64. Glover, P, "Formation Evaluation MSc Course Notes," Aberdeen University, 54–75, 2001.

65. Rahli, O., Tadrist, L., Miscevic, M., Santini, R., "Fluid flow through randomly packed monodisperse fibers: The Kozeny-Carman parameter analysis," Journal of fluids engineering, Vol. 119, No. 1, 188–192, 1997.

66. Rouquerol, J., Baron, G., Denoyel, R., Giesche, H., Groen, J., Klobes, P., Unger, K., "Liquid intrusion and alternative methods for the characterization of macroporous materials (IUPAC Technical Report)," Pure and Applied Chemistry, Vol. 84, No. 1, 107–136, 2011.

67. Cho, D., Naydich, A., Frey, M. W., Joo, Y. L. "Further improvement of air filtration efficiency of cellulose filters coated with nanofibers via inclusion of electrostatically active nanoparticles," Polymer, Vol. 54, No. 9, 2364–2372, 2013.

68. Strange, J. H., Rahman, M., Smith, E. G., "Characterization of porous solids by NMR," Physical review letters, Vol. 71, No. 21, 3589–3591, 1993.

69. Mickel, W., Münster, S., Jawerth, L. M., Vader, D. A., Weitz, D. A., Sheppard, A. P., Schröder-Turk, G. E., "Robust pore size analysis of filamentous networks from three-dimensional confocal microscopy," Biophysical journal, Vol. 95, No. 12, 6072–6080, 2008.

70. Charcosset, C., Cherfi, A., Bernengo, J. C., "Characterization of microporous membrane morphology using confocal scanning laser microscopy," Chemical engineering science, Vol. 55, No. 22, 5351–5358, 2000.

71. Liang, Z. R., Fernandes, C. P., Magnani, F. S., Philippi, P. C., "A reconstruction technique for three-dimensional porous media using image analysis and Fourier transforms," Journal of Petroleum Science and Engineering, Vol. 21, No. 3, 273–283, 1998.

72. Tomba, E., Facco, P., Roso, M., Modesti, M., Bezzo, F., Barolo, M., "Artificial vision system for the automatic measurement of interfiber pore characteristics and fiber

diameter distribution in nanofiber assemblies," Industrial and Engineering Chemistry Research, Vol. 49, No. 6, 2957–2968, 2010.

73. Deshpande, S., Kulkarni, A., Sampath, S., Herman, H., "Application of image analysis for characterization of porosity in thermal spray coatings and correlation with small angle neutron scattering," Surface and Coatings Technology, Vol. 187, No. 1, 6–16, 2004.

74. Pierantonio, F., Masiero, A., Bezzo, F., Beghi, A., Barolo, M., "An Improved Multivariate Image Analysis Method for Quality Control of Nanofiber Membranes," Preprints of the 18th IFAC World Congress Milano (Italy), 2011.

75. Okabe, H., Blunt, M. J., "Pore space reconstruction using multiple-point statistics," Journal of Petroleum Science and Engineering, Vol. 46, No. 1, 121–137, 2005.

76. Levitz, P., "Toolbox for 3D imaging and modeling of porous media: Relationship with transport properties," Cement and concrete research, Vol. 37, No. 3, 351–359, 2007.

77. Jung, Y. J., Baik, A., Kim, J., Park, D., "A novel 2D-to-3D conversion technique based on relative height-depth cue," In IS&T/SPIE Electronic Imaging, International Society for Optics and Photonics, 72371U-72371U, 2009.

78. Ko, J., Kim, M., Kim, C., "2D-to-3D stereoscopic conversion: depth-map estimation in a 2D single-view image," Optical Engineering+ Applications, International Society for Optics and Photonics, 66962A-66962A, 2007.

79. Saxena, A., Chung, S. H., Ng, A. Y., "3-d depth reconstruction from a single still image," International Journal of Computer Vision, Vol. 76, No. 1, 53–69, 2008.

80. Liu, B., Gould, S., Koller, D., "Single image depth estimation from predicted semantic labels," In Computer Vision and Pattern Recognition (CVPR), IEEE Conference, 1253–1260, 2010.

81. Saxena, A., Chung, S. H., Ng, A. Y., "Learning depth from single monocular images," In Advances in Neural Information Processing Systems, 1161–1168, 2005.

82. Kushal, A., Bansal, V., Banerjee, S., "A Simple Method for Interactive 3D Reconstruction, Camera Calibration from a Single View," ICVGIP, 2002.

83. Barinova, O., Konushin, V., Yakubenko, A., Lee, K., Lim, H., Konushin, A., "Fast automatic single-view 3-d reconstruction of urban scenes," Computer Vision–ECCV, Springer Berlin Heidelberg, 100–113, 2008.

84. Srivastava, S., Saxena, A., Theobalt, C., Thrun, S., Ng, A. Y., "i23-Rapid Interactive 3D Reconstruction from a Single Image," VMV, 19–28, 2009.

85. Van den Heuvel, F. A., "3D reconstruction from a single image using geometric constraints," ISPRS Journal of Photogrammetry and Remote Sensing, Vol. 53, No. 6, 354–368, 1998.

86. Ma, Y. (Ed.), "An Invitation to 3D Vision," Springer, LLC, Vol. 26, 2003.

87. Wilson, E. B., "Projective and Metric Geometry," Annals of Mathematics, Second Series, Vol. 5, No. 3, 145–150, 1904.

88. Liebowitz, D., Criminisi, A., Zisserman, A., "Creating Architectural Models from Images," EUROGRAPHICS'99, Vo. 18, No.3, 1999.

89. Germain, S., "Basics of Affine Geometry," Geometric Methods and Applications, 2011.

90. Calabi, E., Olver, P. J., Tannenbaum, A., "Affine geometry, curve flows, and invariant numerical approximations," Advances in Mathematics, Vol. 124, No. 1, 154–196, 1996.

91. Shashua, A., Navab, N., "relative affine structure: theory and application to 3D reconstruction from perspective view," IEEE Computer Society, 1994.

92. Burago, D., Burago, Y., Ivanov, S., "A Course in Metric Geometry," Department of Mathematics, Pennsylvania State University, 2001.

93. Wolpert, S. A., "The Weil-Petersson metric geometry," Handbook of Teichmüller theory, Vol. 2, 47–64, 2009.

94. Havel, T. F., "Some examples of the use of distances as coordinates for Euclidean geometry" Journal of Symbolic Computation, Vol. 11, No. 5, 579–593, 1991.

95. Minkowski, H., "Basics of Euclidean Geometry," Geometric Methods and Applications, 2011.

96. Devernay, F., Faugeras, O., "From Projective to Euclidean Reconstruction," International Conference on Computer Vision and Pattern Recognition, 1996.

97. Bazin, J.C., Seo, Y., Pollefeys, M. "Globally optimal line clustering and vanishing point estimation in manhattan world," CVPR, 2012.

98. Zhang, L., Koch, R., "Vanishing Points Estimation and Line Classification in a Manhattan World," Institute of Computer Science, University of Kiel, Germany, 2012.

99. Beier, T., "Feature-Based Image Metamorphosis," Computer Graphics, Vol. 26, No. 2, 1992.

100. Dollár, P., Tu, Z., Tao, H., Belongie, S., "Feature Mining for Image Classification," CVPR'07, IEEE Conference, 1–8, 2007.

101. Zhou, X. S., Huang, T. S., "Image retrieval: feature primitives, feature representation, and relevance feedback," Content-based Access of Image and Video Libraries, Proceedings, IEEE Workshop, 10–14, 2000.

102. Haralick, R. M., Shanmugam, K., Dinstein, I. H., "Textural features for Image Classification," IEEE Transactions on systems, Man and cybernetics, Vol. SMC-3, No. 6, 610–621,1973.

103. McGuire, M., Hughes, J. F., "Hardware-determined feature edges," Proceedings of the 3rd international symposium on Non-photorealistic animation and rendering, ACM, 35–47, 2004.

104. Harris, C., Stephens, M., "A combined corner and edge detector," Alvey vision conference, Vol. 15, 147–152. 1988.

105. Yamakawa, S., Shimada, K., "Polygon crawling: Feature-edge extraction from a general polygonal surface for mesh generation," Proceedings of the 14th International Meshing Roundtable, Springer Berlin Heidelberg, 257–274, 2005.

106. Azad, P., Asfour, T., Dillmann, R. "Combining Harris interest points and the SIFT descriptor for fast scale-invariant object recognition," Intelligent Robots and Systems, IROS, IEEE/RSJ International Conference, 4275–4280, 2009.

107. Yu, T., Woodford, O. J., Cipolla, R., "An evaluation of volumetric interest points," 3D Imaging, Modeling, Processing, Visualization and Transmission (3DIMPVT), International Conference IEEE, 282–289, 2011.

108. Ferraz, L., Binefa, X., "A scale invariant interest point detector for discriminative blob detection," Universitat Autonoma de Barcelona, Department of Computing Science, Barcelona, Spain, 2009.

109. Manjunath, B. S., "A New Approach to Image Feature Detection with Applications," Department of Electrical and Computer Engineering, University of California, Santa Barbara, 1994.

110. Li, Q., Wang, G., Liu, J., Member, Chen, S., "Robust Scale-Invariant Feature Matching for Remote Sensing Image Registration," IEEE Geoscience And Remote Sensing Letters, Vol. 6, No. 2, 2009.
111. Gurbuz, A. C., "Feature Detection Algorithms in Computed Images," PhD Thesis, Georgia Institute of Technology, 2008.
112. Mikolajczyk, K., Schmid, C., "A performance evaluation of local descriptors," Pattern Analysis and Machine Intelligence, IEEE Transactions, Vol. 27, No. 10, 1615–1630, 2005.
113. Moreno, P., Bernardino, A., Santos-Victor, J., "Improving the SIFT descriptor with smooth derivative filters," Pattern Recognition Letters, Vol. 30, No. 1, 18–26, 2009.
114. Azad, P., Asfour, T., Dillmann, R., "Combining Harris interest points and the SIFT descriptor for fast scale-invariant object recognition," Intelligent Robots and Systems, IEEE/RSJ International Conference, 4275–4280, 2009.
115. Grauman. K, Leibe. B, "visual recognition: Local Features: Detection and Description," 23–39.
116. Bhatia, A., "Hessian-Laplace feature detector and Haar descriptor for image matching," MSc thesis, Doctoral dissertation, University of Ottawa, 2007.
117. Remondino, F., "Detectors and descriptors for photogrammetric applications," International Archives of Photogrammetry, Remote Sensing and Spatial Information Sciences, Vol. 36, No. 3, 49–54, 2006.
118. Laptev, I., "On space-time interest points," International Journal of Computer Vision, Vol. 64, No. 2–3, 107–123, 2005.
119. Carmichael, G., Laganière, R., Bose, P., "Global Context Descriptors for SURF and MSER Feature Descriptors," Computer and Robot Vision (CRV), Canadian Conference IEEE, 309–316, 2010.
120. Forssen, P. E., Lowe, D. G., "Shape descriptors for maximally stable extremal regions," Computer Vision, ICCV, IEEE 11th International Conference, 1–8, 2007.
121. Chien, H. C., "On the evaluation of interest point detectors algorithm by using satellite images," Doctoral dissertation, The Ohio State University, 2010.
122. Rodehorst, V., Koschan, A., "Comparison and evaluation of feature point detectors," In Proc. 5th International Symposium Turkish-German Joint Geodetic Days Geodesy and Geoinformation in the Service of our Daily Life," Berlin, Germany, 2006.
123. Mortensen, E. N., Deng, H., Shapiro, L., "A SIFT descriptor with global context," Computer Vision and Pattern Recognition (CVPR), IEEE Computer Society Conference, Vol. 1, 184–190, 2005.
124. Läbe, T., Förstner, W., "Automatic relative orientation of images," Proceedings of the 5th Turkish-German Joint Geodetic Days, Vol. 29, No. 31, 2006.
125. Scovanner, P., Ali, S., Shah, M., "A 3-dimensional sift descriptor and its application to action recognition," In Proceedings of the 15th international conference on Multimedia, 357–360, ACM, 2007.
126. Grabner, M., Grabner, H., Bischof, H., "Fast approximated SIFT," Computer Vision–ACCV, Springer Berlin Heidelberg, 918–927, 2006.
127. Boureau, Y. L., "Learning Hierarchical Feature Extractors For Image Recognition," PhD Thesis. New York Univ Ny Dept of Computer Science, 2012.

128. Ke, Y., Sukthankar, R., "PCA-SIFT: A more distinctive representation for local im-age descriptors," Computer Vision and Pattern Recognition, CVPR, Proceedings of the IEEE Computer Society Conference, Vol. 2, II-506, 2004.
129. Lowe, D. G., "Object recognition from local scale-invariant features," Computer vi-sion, the proceedings of the seventh IEEE international conference, Vol. 2, 1150–1157, 1999.
130. Abdel-Hakim, A. E., Farag, A. A., "CSIFT: A SIFT descriptor with color invariant characteristics," Computer Vision and Pattern Recognition, IEEE Computer Society Conference, Vol. 2, 1978–1983, 2006.
131. Thomee, B., Bakker, E. M., Lew, M. S., "TOP-SURF: a visual words toolkit," Pro-ceedings of the international conference on Multimedia, 1473–1476, ACM, 2010.
132. Ballesta, M., Gil, A., Mozos, O. M., Reinoso, O., "Local descriptors for visual SLAM," Workshop on Robotics and Mathematics (ROBOMAT07), Portugal, 209–215, 2007.
133. Bay, H., Tuytelaars, T., Van Gool, L., "Surf: Speeded up robust features," Computer Vision–ECCV, Springer Berlin Heidelberg, 404–417, 2006.
134. Kim, D., Dahyot, R., "Face components detection using SURF descriptors and SVMs," Machine Vision and Image Processing Conference, IMVIP'08, IEEE Inter-national, 51–56, 2008.
135. Rublee, E., Rabaud, V., Konolige, K., Bradski, G., "ORB: an efficient alternative to SIFT or SURF," Computer Vision (ICCV), IEEE International Conference, 2564–2571, 2011.
136. Bay, H., Ess, A., Tuytelaars, T., Van Gool, L., "Speeded-up robust features (SURF)," Computer vision and image understanding, Vol. 110, No. 3, 346–359, 2008.
137. Dalal, N., Triggs, B., "Histograms of oriented gradients for human detection," Com-puter Vision and Pattern Recognition, IEEE Computer Society Conference (CVPR), Vol. 1, 886–893, 2005.
138. Doersch, C., Efros, A., "Improving the HoG Descriptor," http://www.cs.cmu.edu/-cdoersch/projects/hogimprove, 2012.
139. Hu, R., Barnard, M., Collomosse, J., "Gradient field descriptor for sketch based re-trieval and localization," Image Processing (ICIP), 17th IEEE International Confer-ence, 1025–1028, 2010.
140. Suard, F., Rakotomamonjy, A., Bensrhair, A., Broggi, A., "Pedestrian detection using infrared images and histograms of oriented gradients," Intelligent Vehicles Sympo-sium, IEEE, 206–212, 2006.
141. Chen, F., Yu, S. N., "Content-based image retrieval by DTCWT feature," Computer Research and Development (ICCRD), 3rd International Conference, IEEE, Vol. 4, 283–286, 2011.
142. Nelson, J. D., Pang, S. K., Kingsbury, N. G., Godsill, S. J., "Tracking ground based targets in aerial video with dual-tree wavelet polar matching and particle filtering," FUSION, 1321–13277, 2008.
143. Khanapuri, J., Kulkarni, L. "A Novel Approach for Color Image Retrieval Using Complex Wavelets and Color Descriptors," 2013.
144. Liu, Y., Zhou, X., "A simple texture descriptor for texture retrieval," Communication Technology Proceedings, ICCT, International Conference IEEE, Vol. 2, 1662–1665, 2003.

145. George, J. P., Abhilash, S. K., Raja, K. B., "Transform Domain Fingerprint Identification Based on DTCWT," International Journal of Advanced Computer Science and Applications, Vol. 3, No. 1, 2012.

146. Murillo, A. C., Kosecka, J., "Experiments in place recognition using gist panoramas," Computer Vision Workshops (ICCV Workshops), IEEE 12th International Conference, 2196–2203, 2009.

147. Torralba, A., Fergus, R., Weiss, Y., "Small codes and large image databases for recognition," Computer Vision and Pattern Recognition (CVPR), IEEE Conference, 1–8, IEEE, 2008.

148. Murillo, A. C., Campos, P., Kosecka, J., Guerrero, J. J., "Gist vocabularies in omnidirectional images for appearance based mapping and localization," 10th OMNIVIS, held with Robotics: Science and Systems (RSS), Vol. 3, 2010.

149. Hays, J., Efros, A. A., "Scene completion using millions of photographs," ACM Transactions on Graphics (TOG), Vol. 26, No. 3, p. 4, 2007.

150. Sikirić, I., Brkić, K., Šegvić, S., "Classifying traffic scenes using the GIST image descriptor," Proceedings of the Croatian Computer Vision Workshop, 2013.

151. Torralba, A., Murphy, K. P., Freeman, W. T., "Using the forest to see the trees: exploiting context for visual object detection and localization," Communications of the ACM, Vol. 53, No. 3, 107–114, 2010.

152. Douze, M., Jégou, H., Sandhawalia, H., Amsaleg, L., Schmid, C., "Evaluation of GIST descriptor s for web-scale image search," Proceedings of the ACM International Conference on Image and Video Retrieval, 19, 2009.

153. Levi, D., Silberstein, S., Bar-Hillel, A., "Fast multiple-part based object detection using KD-Ferns," Computer Vision and Pattern Recognition (CVPR), IEEE Conference, 947–954, 2013.

154. Wagner, D., Reitmayr, G., Mulloni, A., Drummond, T., Schmalstieg, D., "Real-time detection and tracking for augmented reality on mobile phones," Visualization and Computer Graphics, IEEE Transactions, Vol. 16, No. 3, 355–368, 2010.

155. Bosch, A., Zisserman, A., Munoz, X., "Image classification using random forests and ferns," 2007.

156. Oshin, O., Gilbert, A., Illingworth, J., Bowden, R., "Spatio-temporal feature recognition using randomized ferns," The 1st International Workshop on Machine Learning for Vision-based Motion Analysis-MLVMA'08, 2008.

157. Ozuysal, M., Calonder, M., Lepetit, V., Fua, P. "Fast keypoint recognition using random ferns," Pattern Analysis and Machine Intelligence, IEEE Transactions, Vol. 32, No. 3, 448–461, 2010.

158. Wagner, D., Reitmayr, G., Mulloni, A., Drummond, T., Schmalstieg, D., "Pose tracking from natural features on mobile phones," Proceedings of the 7th IEEE/ACM International Symposium on Mixed and Augmented Reality, IEEE Computer Society, 125–134, 2008.

159. Hinterstoisser, S., Kutter, O., Navab, N., Fua, P., Lepetit, V., "Real-time learning of accurate patch rectification," Computer Vision and Pattern Recognition (CVPR), IEEE Conference, 2945–2952, 2009.

160. Dahl, A. L., Aanæs, H., Pedersen, K. S., "Finding the best feature detector-descriptor combination," 3D Imaging, Modeling, Processing, Visualization and Transmission (3DIMPVT), IEEE International Conference, 318–325, 2011.

161. Winder, S., Hua, G., Brown, M., "Picking the best daisy," Computer Vision and Pattern Recognition (CVPR), IEEE Conference, 178–185, 2009.
162. Fischer, J., Ruppel, A., Weisshardt, F., Verl, A., "A rotation invariant feature descriptor O-DAISY and its FPGA implementation," Intelligent Robots and Systems (IROS), IEEE/RSJ International Conference, 2365–2370, 2011
163. Candocia, F., Adjouadi, M., "A Similarity Measure for Stereo Feature Matching," IEEE Transactions On Image Processing, Vol. 6, No. 10, 1997.
164. Baumberg, A., "Reliable Feature Matching Across Widely Separated Views," Computer Vision and Pattern Recognition, 2000.
165. Brown, M., Lowe, D. G., "Unsupervised 3D Object Recognition and Reconstruction in Unordered Datasets," Department of Computer Science, University of British Columbia, 2006.
166. Lu, N., Feng, Z., "Mathematical model of blob matching and modified Bhattacharyya coefficient," Image and Vision Computing, Vol. 26, No. 10, 1421–1434, 2008.
167. Delponte, E., Isgrò, F., Odone, F., Verri, A., "SVD-matching using SIFT features," Graphical models, Vol. 68, No. 5, 415–431, 2006.
168. Mori, G., Belongie, S., Malik, J., "Shape contexts enable efficient retrieval of similar shapes," In Computer Vision and Pattern Recognition (CVPR), Proceedings of the IEEE Computer Society Conference, Vol. 1, I-723, 2001.
169. Awad, G., Han, J., Sutherland, A., "A unified system for segmentation and tracking of face and hands in sign language recognition," Pattern Recognition (ICPR), 18th International Conference IEEE, Vol. 1, 239–242, 2006.
170. Kong, S., Sanderson, C., Lovell, B. C., "Classifying and tracking multiple persons for proactive surveillance of mass transport systems," Advanced Video and Signal Based Surveillance (AVSS), IEEE Conference, 159–163, 2007.
171. Foresti, G. L., "Real time detection of multiple moving objects in complex image sequences," International journal of imaging systems and technology, Vol. 10, No. 4, 305–317, 1999.
172. Lalonde, M., Beaulieu, M., Gagnon, L., "Fast and robust optic disc detection using pyramidal decomposition and Hausdorff-based template matching," Medical Imaging, IEEE Transactions, Vol. 20, No. 11, 1193–1200, 2001.
173. Slaney, J., Fujita, M., Stickel, M., "Automated reasoning and exhaustive search: Quasigroup existence problems," Computers & mathematics with applications, Vol. 29, No. 2, 115–132, 1995.
174. Mertens, S., "Exhaustive search for low-autocorrelation binary sequences," Journal of Physics A: Mathematical and General, Vol. 29, No. 18, L473, 1996.
175. Berg, A. C., Malik, J., "Geometric blur for template matching," Computer Vision and Pattern Recognition (CVPR), Proceedings of the IEEE Computer Society Conference, Vol. 1, I-607, 2001.
176. Barni, M., "Effectiveness of exhaustive search and template matching against watermark desynchronization," Signal Processing Letters, IEEE, Vol. 12, No. 2, 158–161, 2005.
177. Lewis, J. P., "Fast template matching," Vision interface, Vol. 95, No. 120123, 15–19, 1995.
178. Wu, Q., Yu, Y., "Feature matching and deformation for texture synthesis," ACM Transactions on Graphics (TOG), Vol. 23, No. 3, 364–367, 2004.

179. Zhou, H., Huang, T., "Okapi-chamfer matching for articulate object recognition," Computer Vision, ICCV, Tenth IEEE International Conference, Vol. 2, 1026–1033, 2005.

180. Liu, M. Y., Tuzel, O., Veeraraghavan, A., Chellappa, R., "Fast directional chamfer matching," Computer Vision and Pattern Recognition (CVPR), IEEE Conference, 1696–1703, 2010.

181. Park, S. C., Lim, S. H., Sin, B. K., Lee, S. W., "Tracking nonrigid objects using probabilistic Hausdorff distance matching," Pattern Recognition, Vol. 38, No. 12, 2373–2384, 2005.

182. Kwon, O. K., Sim, D. G., Park, R. H., "Robust Hausdorff distance matching algorithms using pyramidal structures," Pattern Recognition, Vol. 34, No. 10, 2005–2013, 2001.

183. Sim, D. G., Kwon, O. K., Park, R. H., "Object matching algorithms using robust Hausdorff distance measures," IEEE Transactions on Image Processing, Vol. 8, No. 3, 425–429, 1999.

184. Zhang, D., Lu, G., "Review of shape representation and description techniques," Pattern recognition, Vol. 37, No. 1, 1–19, 2004.

185. Nuchter, A., Lingemann, K., Hertzberg, J., "Cached kd tree search for ICP algorithms," 3D Digital Imaging and Modeling (3DIM'07), Sixth IEEE International Conference, 419–426, 2007.

186. Zhou, K., Hou, Q., Wang, R., Guo, B., "Real-time KD-tree construction on graphics hardware," ACM Transactions on Graphics (TOG), Vol. 27, No. 5, 126, 2008.

187. Beis, J. S., Lowe, D. G., "Shape indexing using approximate nearest-neighbor search in high-dimensional spaces," Computer Vision and Pattern Recognition, IEEE Computer Society Conference, 1000–1006, 1997.

188. Foley, T., Sugerman, J., "KD-tree acceleration structures for a GPU raytracer," Proceedings of the ACM SIGGRAPH/EUROGRAPHICS conference on Graphics hardware, 15–22, 2005.

189. Muja, M., Lowe, D. G., "Fast Approximate Nearest Neighbors with Automatic Algorithm Configuration," VISAPP, Vol. 1, 331–340, 2009.

190. Lowe, D. G., "Object recognition from local scale-invariant features," Computer vision, The proceedings of the seventh IEEE international conference, Vol. 2, 1150–1157, 1999.

191. Lowe, D. G., "Distinctive Image Features from Scale-Invariant Keypoints," Computer Science Department University of British Columbia, 2004.

192. Heikkila, J., "Geometric camera calibration Using circular control points," IEEE Transactions on pattern analysis and machine intelligence, Vol. 22, No. 10, 2000.

193. Mi˘cu˘s'ık, B., Pajdla, T., "Estimation of omnidirectional camera model from epipolar geometry," Proceedings of Conference on Computer Vision and Pattern Recognition, IEEE Computer Society, Madison, Wisconsin, USA, 2003.

194. Pollefeys, M., Van Gool, L., Proesmans, M., "Euclidean 3D reconstruction from image sequences with variable focal lengths," Computer Vision—ECCV'96, 1996.

195. Loop, C., Zhang, Z., "Computing Rectifying Homographies for Stereo Vision," IEEE Conference on Computer Vision and Pattern Recognition, Vol. 1, 125–131, 1999.

196. Fusiello, A., Trucco, E., Verri, A., "A compact algorithm for rectification of stereo pairs," Machine Vision and Applications, Vol. 12, 16–22, 2000.

197. Oram, D., "Rectification for Any Epipolar Geometry," BMVC, comp.leeds.ac.uk, 2001.
198. Pollefeys, M., Koch, R., Van Gool, L., "A simple and efficient rectification method for general motion," Computer Vision, 1999.
199. Sandr, J., "Epipolar Rectification for Stereovision," Research Reports of CMP, Czech Technical University in Prague, No. 4, 2009.
200. Dubrofsky, E., "Homography Estimation," MSC thesis, The University of British Columbia, 2009.
201. Zhang, Z., Hanson, AR., "Scaled Euclidean 3D reconstruction based on externally uncalibrated cameras," Browse Conference Publications, Computer Vision, 1995.
202. Zhang, Z., Hanson, AR., "3D reconstruction based on homography mapping," Proc. ARPA96, 1007–1012, 1996.
203. Trung Kien, D., "A Review of 3D Reconstruction from Video Sequences," University of Amsterdam ISIS Technical Report Series, 2005.
204. Barinova, O., Konushin, V., Yakubenko, A., Lee, K., Lim, H., Konushin, A., "Fast automatic single-view 3D reconstruction of urban images," Computer Vision – ECCV 2008, Lecture Notes in Computer Science, Vol. 5303, 100–113, 2008.
205. Mouragnon, E., Maxime L., Michel D., Fabien D., Patrick S., "Real time localization and 3d reconstruction," Computer Vision and Pattern Recognition, IEEE Computer Society Conference, Vol. 1, 363–370, 2006.
206. Triggs, B., McLauchlan, P. F., Hartley, R. I., Fitzgibbon, A. W., "Bundle adjustment—a modern synthesis," Vision algorithms: theory and practice, Springer Berlin Heidelberg, 298–372, 2000.
207. Engels. C., Stewénius. H., Nistér. D., "Bundle adjustment rules," Photogrammetric computer vision, Vol. 2, 2006.
208. Rachmielowski, A., Birkbeck, N., Jagersand, M., Cobzas, D., "Realtime visualization of monocular data for 3D reconstruction," Computer and Robot Vision (CRV'08), Canadian Conference on IEEE, 196–202, 2008.
209. Ioannides, M., Stylianidis, E., Stylianou, S., "3D reconstruction and visualization in cultural heritage," Proceedings of the 19th International CIPA Symposium, 258–262, 2003.
210. Mayer. H., Neubiberg, "3D reconstruction and visualization of urban images from uncalibrated wide-baseline image sequences," Photogramtrie Fernerkundung Geoinformation, Vol. 167, No. 3, 2007.
211. Haala, N., sttutgart, "Multiray photogrammetry and dense image matching" Photogramtrische Woche, 2011.
212. Wei, Y., Lhuillier, M., Quan, L., "Fast segmentation-based dense stereo from quasi-dense matching," Asian Conference on Computer Vision, 2004.
213. Braux-Zin, J., Dupont, R., Bartoli, A., "A General Dense Image Matching Framework Combining Direct and Feature-Based Costs," Computer Vision (ICCV), IEEE International Conference, 185–192, 2013.
214. Mathias, R., Haala, N., "Potential of Dense Matching for the Generation of High Quality Digital Elevation models" In Proceedings of ISPRS Hannover Workshop High-Resolution Earth Imaging for Geospatial Information, 331–343. 2011.
215. Jaechul, K., Liu, C., Sha, F., Grauman, K., "Deformable spatial pyramid matching for fast dense correspondences" In Computer Vision and Pattern Recognition (CVPR), IEEE Conference, 2307–2314, 2013.

216. Leordeanu, M., Zanfir, A., Sminchisescu, C., "Locally affine sparse-to-dense matching for motion and occlusion estimation," Computer Vision (ICCV), IEEE International Conference, 1721–1728, 2013.
217. Megyesi, Z., "Dense Matching Methods for 3D Image Reconstruction from Wide Baseline Images," PhD Thesis, France, Eotvos Lorand University, 2009.
218. Haala, N., "The Landscape of Dense Image Matching Algorithms," 2013.
219. Bartoli, A., Sturm, P., "Structure-from-motion using lines: Representation, triangulation, and bundle adjustment," Computer Vision and Image Understanding, Vol. 100, No. 3, 416–441, 2005.
220. Deok-Soo, K., Donguk, K., Youngsong, C., Kokichi S., "Quasi-triangulation and interworld data structure in three dimensions," Computer-Aided Design, Vol. 38, No. 7, 808–819, 2006.
221. Lee, D. T., Schachter, B. J., "Two algorithms for constructing a Delaunay triangulation," International Journal of Computer and Information Sciences, Vol. 9, No. 3, 219–242, 1980.
222. Hartle, R. I., "Triangulation," Computer Vision and Image Understanding, Vol. 68, No. 2, 146–157, 1997.
223. Kirby, R. C., Siebenmann, L. C., "On the triangulation of manifolds and the Hauptvermutung," Bull. Amer. Math. Soc, Vol. 75, 742–749, 1969.
224. Aurenhammer, F., "Voronoi diagrams—a survey of a fundamental geometric data structure," ACM Computing Surveys (CSUR), Vol. 23.No. 3, 345–405, 1991.
225. Heckbert, P. S., "Fundamentals of texture mapping and image warping," MSc thesis, University of California, Berkeley, 1989.
226. Haeberli, P., Segal, M., "Texture mapping as a fundamental drawing primitive" In Fourth Eurographics Workshop on Rendering, Vol. 259, 266, 1993.
227. Wiederkehr, T., Klusemann, B., Gies, D., Müller, H., Svendsen, B., "An image morphing method for 3D reconstruction and FE-analysis of pore networks in thermal spray coatings," Computational Materials Science, Vol. 47, 881–889, 2010.
228. Delerue, J. F., Perrie, E., Yu, Z. Y., Velde, B., "New Algorithms in 3D Image Analysis and their Application to the Measurement of a Specialized Pore Size Distribution in Soils," Phys. Chem. Earth(A), Vol. 24, No. 7, 639–644, 1999.
229. Al-Raoush, R.I., Willson, C.S., "Extraction of physically realistic pore network properties from three-dimensional synchrotron X-ray microtomography images of unconsolidated porous media systems," Journal of Hydrology, Vol. 300, 44–64, 2005.
230. Liang, Z. R., Fernandes, C.P., Magnani, F.S., Philippi, P.C., "A reconstruction technique for three-dimensional porous media using image analysis and Fourier transforms," Journal of Petroleum Science and Engineering Vol. 21, 273–283,1998.
231. Diógenes, A. N., dos Santos, L. O. E., Fernandes, C. P., Moreira, A. C., Apolloni, C. R., "Porous Media Microstructure Reconstruction Using Pixel-Based And Object-Based Simulated Annealing – Comparison With other Reconstruction Methods," Engenharia Térmica (Thermal Engineering), Vol. 8, No. 02, 35–41, 2009.
232. Faessel, M., Delisee, C., Bos, F., Castera, P., "3D Modeling of random cellulosic fibrous networks based on X-ray tomography and image analysis," Composites Science and Technology, Vol. 65, 1931–1940, 2005.

CHAPTER 7

STRUCTURE AND OZONE RESISTANCE OF VULCANIZED ACRYLONITRILE-BUTADIENE RUBBERS WITH POLY(VINYL CHLORIDE) BLENDS

N. M. LIVANOVA, A. A. POPOV, S. G. KARPOVA, and G. E. ZAIKOV

Emanuel Institute of Biochemical Physics, Russian Academy of Sciences, 4 Kosygina Str., 119991 Moscow, Russia; E-Mail: Livanova@Sky.chph.ras.ru

CONTENTS

ABSTRACT

The ozone resistance, phase structure, and the structure of interphase transition layers in vulcanized blends of butadiene-acrylonitrile rubbers (NBR) with poly(vinyl chloride) (PVC) was studied by measuring mechanical stress relaxation in an inert and ozone-containing atmosphere and EPR. The samples were prepared from NBR of the SKN-18, SKN-26, and SKN-40 grades and PVC (obtained by emulsion or suspension polymerization) by methods of high- and low-temperature blending followed by treatment in a thermostat. A mechanism of the ozone-protective action of thermoplastic (PVC) in the blends with NBR is considered for the systems possessing various degrees of homogeneity.

7.1   INTRODUCTION

The problem of increasing the stability of unsaturated rubbers [in particular, of the butadiene-acrylonitrile rubbers (NBR)] with respect to ozone by blending them with saturated polymers was extensively studied [1–5]. Only a small proportion of the saturated polymers are capable of providing the ozone protection [PVC, perchloro vinyl polymer, ethylene-propylene diene terpolymer (EPDE), ethylene-propylene rubber (EPR)] [1–7]. There is no commonly accepted opinion concerning the mechanism of the ozone-protective action. In particular, Zateev [6] showed that effective protection is provided only if the polymer forms a continuous phase and is highly homogenized. Khanin [7] explained the protective properties by the formation of a surface layer in elastomer enriched with the saturated polymer. It was established [1–5] that necessary conditions for the obtaining of materials with high ozone resistance include a sufficiently high degree of homogeneity and the formation of a continuous framework of the ozone-resistant component (achieved at a 30% PVC content in the blend).

In the works of Refs. [4, 5], it was studied that the relation of ozone resistance to the volume and structure of the interfacial layer and the amounts of crosslinks in the interlayer for co-vulcanizates of acrylonitrile-butadiene rubbers of various polarities with ethylene–propylene–diene (EPDM) elastomers that differed in the co-monomer composition and stereoregularity of propylene units. It was shown that the ozone resistance is determined by the compatibility of the components, phase structure, the interlayer volume and density, the amount of crosslinks in the interlayer, and the strength of the EPDM network.

As the content of nitrile groups in NBR grows, the polarity of the rubber and, hence, its compatibility with polar PVC, tend to increase

[1–3, 8–16]. However, this is accompanied by the loss of elasticity, a drop in the cold stability, and a decrease in the electrical properties of vulcanizates [1]. In this connection, there are many works devoted to attempts made to improve the miscibility of PVC with rubbers having medium contents of nitrile groups. Experiments using various blending techniques (low- and high-temperature mechanical blending, latex coagulation, coprecipitation from solution) [10–16] showed that the final blends may acquire, depending on the blending method used, the properties of both a homo- and heterogeneous system. The mechanical blending typically leads to nonequilibrium systems, whereby the two phases formed by each of the components are separated by a transition layer formed as a result of mutual diffusion. This layer is manifested in thermograms [10, 11], NMR spectra [12, 13], and the temperature dependence of the mechanical loss tangent [14].

The purpose of this work was to establish a relationship between morphology of the interphase contact region and the ozone resistance of vulcanized blends and to refine the above mechanism of the ozone-protective thermoplast action in the systems of components with various degrees of compatibility and homogeneity. Likewise we have performed this work in order to establish the most important structural and technological parameters affecting stability of the PVC-NBR system with respect to ozone. For this purpose, we have varied polarity of the NBR matrix by changing the content of nitrile groups and PVC (obtained by emulsion or suspension polymerization) and employed the technology of low- temperature blending followed by high-temperature treatment of the blend to facilitate the mutual diffusion of components at the phase boundaries and high-temperature mechanical blending.

7.2 EXPERIMENTAL

The experiments were performed using vulcanized blends of NBR (commercial SKN-18, SKN-26, and SKN-40 grades containing 18, 26, and 40% of acrylonitrile units) and PVC (obtained by emulsion or suspension polymerization—PVC-E and PVC-S, respectively) with the 70:30 ratio of components. The rubbers with 26–28% content of acrylonitrile groups were represented by Paracril BJ (Uniroyal Co., USA) containing

28% of acrylonitrile units (series I), SKN-26 (series II), and SKN-26 M (series III). Paracril BJ is characterized by the Mooney viscosity 50 ± 7 at 118°C. The samples of SKN-26 and SKN-26 M had M_n =(1–3) × 10^5, the Mooney viscosities above 70 and 55, and the Defo harnesses about 20 and 10 N, respectively. PVC-E was represented by a commercial PVKh-E- 6250-Zh grade with M_n = (7.0–7.5) × 10^4. The PVC-S samples were as follows: PVC VISTA 5305 grade (USA)—a GAS 980-Z-86-2 homopolymer with an intrinsic viscosity of 0.72 (series I) and PVC S-7058 M grade with M_n = 6.4 × 10^4, M_w = 2 × 10^5, and a specific viscosity in dichloroethane of 1.31 (series II and III).

SKN-26 and SKN-40 were blended with PVC-E at 40°C as described in Ref. [17], followed by treatment in a thermostat at 120, 140, or 160°C (series A). Analogous procedures were used to obtain the blends of NBR with PVC-S (series B). The high-temperature blending of NBR samples containing 26–28% of acrylonitrile (AN) with PVC-S was performed by 8-min treatment at 170°C in a mixer operated at 60 rpm (series C). The treatment was performed in the presence of thermostabilizing agents: an epoxidated soybean oil Paraflex G-62 (3 wt fractions) for Paracril BJ (series I) or an ED-20 epoxide resin (2 wt fractions) for SKN-26 (series II) and SKN-26 M (series III). For the comparison, one sample was prepared by blending SKN-26 with PVC in the same mixer operated at a reduced speed of 30 rpm. The blending with vulcanizing system was performed according to a conventional roller technology for 20 min at 40°C, and the vulcanization was effected for 20 min at 165°C. The vulcanizing systems in series A and B, C were as follows (wt fractions): thiuram, 1.2 (A) and 1.4 (B, C); sulfenamide C, 1.2 (A) and 1.4 (B, C); stearic acid, 0.6 (A) and 0.7 (B, C); zinc oxide, 3.0 (A) and 3.5 (B, C); calcium stearate, 2.0 (A) and 0 (B, C).

The structure and ozone resistance of vulcanized blend samples were studied by measuring the rate of stress relaxation in air (v_{ph}, physical relaxation) and in an ozone-air mixture (v) with the ozone concentration 10^{-5} mol/L. The relaxation measurements were performed on an IKhF-2 relaxometer [17, 18].

The temperature dependence of the relaxation rate was studied in the range from 303 to 363 K on 30%-strained samples. This deformation level is close to a critical value, for which a maximum rate of the ozone induced degradation of rubbers is observed [19, 20]. The stress relaxation rate as function of the sample deformation level was studied at 303 K. The samples had the form of 0.2–0.4-mm-thick plates. The experimental error of measurement of the relaxation rate was ±10%. The relaxation rates were averaged over the results of five independent determinations.

The dynamics of molecules in the blend compositions was studied by the EPR spin-probe technique, using 2,2,6,6-tetramethyl-1-piperidinyloxy (TEMPO, radical 1) and TEMPO-4-benzoate (radical 2) as the spin probes. Radical 1 was introduced into vulcanized samples from the vapor phase, by exposure for 1–2 days at room temperature, followed by equilibration for 1–2 weeks. Radical 2 was introduced into the samples from an acetone solution, followed by the solvent removal in vacuum at 100°C. The radical concentration in the samples was 10^{-4} mol/L. The X-band EPR spectra were measured using an EPR-V spectrometer. The rotational mobility of nitroxyl radicals was characterized by the corresponding correlation time.

7.3 RESULTS AND DISCUSSION

The oxidation process is initiated at the surface. Since the rate of ozone diffusion in polymers is extremely small [19], this oxidizing agent can penetrate deep into a sample only via cavities, pores, voids, and cracks. Mechanism inhibiting the ozone-induced degradation of unsaturated rubbers must be operative in their blends with PVC can be related to a phase structure of the blend and certain morphological features in the region of contact between the blend components.

In the previous works [4, 5, 17], we have demonstrated a relationship between the ozone-protective effect of a thermoplastic component and the structure of an interphase transition layer [17, 21, 22]. The protective effect of PVC was explained by termination of the growth and coalescence of the ozone-induced microcracks in the interphase transition layer. The protective effect decreases with increasing temperature and deformation of a blend because of activation of the segmental mobility in NBR and PVC [17] as a result of the microphase separation occurring in the transition layers—the weakest structural elements of the strained material.

Thus, a high energy of interphase adhesion in blends of compatible components is a factor that imparts strength to the transition layer and renders a vulcanized blend resistant with respect to the ozone-induced cracking.

The compatibility of NBR and PVC depends on the content of polar AN units in the rubber. Rubbers containing 26 and 40% of AN units (SKN-26 and SKN-40) are compatible with PVC. A lower content of AN units (SKN-18) results in a limited miscibility of components, and these blends exhibit very low ozone resistance [1–3]. The ozone-protective thermoplast effect depends on the degree of blend homogeneity [1–3, 17].

The homogeneity of NBR-PVC blends depends on the method of thermoplast preparation and drops sharply on the passage from PVC-E to PVC-S. This is explained by differences in the size, structure, and density of PVC particles and the amount and composition of impurities [23, 24]. A homogeneous blend of NBR with PVC-S can be obtained by high-temperature blending in a mixer [25].

The compatibility and phase structure of NBR-PVC blends were studied by various methods [10–16]. It particular, it was found [12, 14] that the temperature dependence of the mechanical loss tangent of the blends of NBR (with different polarities) and PVC exhibit peaks related to the mechanical vitrification of rubber and plastic components. An additional (intermediate) maximum reflecting the ψ-process of activation of the segmental mobility of macromolecules in the transition layer was reported in Refs. [16, 26].

The temperature of the ψ-transition in a blend exhibits a shift depending on the degree of component compatibility [17] and on the temperature of heat-treatment of the SKN-26 and SKN-40 rubbers blended with PVC-E in series A (Fig. 7.1). The temperature effect was less pronounced for the most compatible blend of SKN-40 (Fig. 7.1b) and a system of limited compatibility based on SKN-18 (Fig. 7.1c). For the blends of SKN-26 (Fig. 7.1a), the maximum activation energies and temperatures of the ψ-transition (and, hence, a maximum density of the interphase transition layer) were observed in the samples treated at 120°C. This regime of blending is optimum for the given system. Treatment at a higher temperature (especially at 160°C) is accompanied by HCl evolution from PVC and the thermal vulcanization of NBR. This results in increasing hardness of the blend and decreasing homogeneity of the component

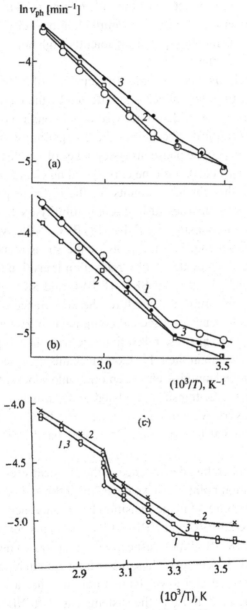

FIGURE 7.1 Temperature dependence of the rate of physical stress relaxation at 30°C in 70: 30 (a) SKN-26-PVC, (b) SKN-40-PVC and SKN-18 with PVC-E blends obtained by low- temperature blending followed by the heat-treatment in a thermostat at $T = 120$ (*1*), 140 (*2*), or 160°C (*3*) (series A, $\varepsilon = 30\%$).

distribution upon introduction of the vulcanizing system, which eventually leads to lower a degree of the sample homogeneity. The absence of other transitions is indicative of a sufficiently high homogeneity (single-phase character) of the blends.

An increase in the compatibility (miscibility) of NBR and PVC with increasing polarity of the rubber is manifested by the increasing homogeneity of the mixture provided otherwise equal conditions (i.e., the same thermostat temperature). The higher the compatibility, the lower is the PVC content at which the phase inversion takes place (due to a higher dispersity of PVC particles), with the corresponding change in the transition layer structure. This change accounts for the different physicomechanical and relaxation properties of high-compatibility (NBR-26 or NBR-40 based) and limited-compatibility (NBR-18) PVC- NBR systems.

As is seen from Fig. 7.1, the plots of the stress relaxation rate versus temperature show evidence of a relaxation transition (at 30–40°C for SKN-26-PVC and 25–30°C for SKN-18-PVC and SKN-40-PVC), which is absent in the pure rubber. Therefore, the appearance of this transition is related to the presence of a second component (PVC) in the blend. An increase in the physical stress relaxation rate with the temperature can be explained by a change in the relaxation mechanism caused by the activation of the segmental mobility chains entering into a looser transition layer. Similar phenomena (ψ-transition), related to the structure of microphases and transition layers, were observed in the NBR-PVC systems by various methods (mechanical loss tangent, DTA, wideband NMR, and acoustic measurements).

Using the temperature dependence of the physical stress relaxation rate above the transition point, we have determined the activation energies for the segmental mobility of in the transition layer. This energy had close values for the NBR-26-PVC and NBR-40-PVC systems containing 20–30% PVC-E (20 and 18–19 kJ/mol, respectively) and was somewhat lower for NBR-18-PVC-E (13 kJ/mol). Thus, the ψ-transition in NBR-18-PVC vulcanizates is observed at a lower temperature and has a lower activation energy as compared to those in the systems based on NBR-26 and NBR-40. This is consistent with the above assumption concerning facilitated increase of the segmental mobility in loose transition layers in comparison to the compatible systems.

We should like to emphasize the coincidence of the temperature interval of the ψ-transition with the published data [10–16]. The NBR-18-PVC system of limited-compatibility shows an additional transition in the region of 60°C, which is not observed in the systems with NBR-26 and NBR-40. This difference can be explained by a higher homogeneity of the latter blends. The transition in the region of 60°C – is λ-relaxations related to formation of fluctuation structures involving *trans*-butadiene units [27, 28].

The experimental data obtained show evidence that the method of stress relaxation rate determination is sensitive to phase transitions and structural changes in the transition layer.

Figure 7.2 shows the plots of the strain at break (ε_b) and the conventional stress at a 300% strain (σ_{300}) determined at room temperature versus the PVC-E content in its blends with NBR-18, NBR-26, and NBR-40 obtained upon the treatment at 120°C. As is seen, the strain at break

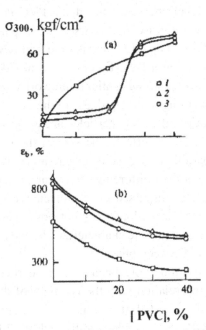

FIGURE 7.2 The plots of (a) conventional stress σ_{300} and (b) strain at break ε_b vs. PVC-E content in vulcanized blends based on (1) NBR-18, (2) NBR-26, and (3) NBR-40 (T=293 K; air).

decreases with increasing content of the rigid-chain polymer in all rubber compositions. After the formation of a more or less (depending on the polarity of rubber) developed spatial network of the PVC-E particles. The ε_b value attains a plateau. As the compatibility of rubber and thermoplast increases, the ε_b plateau is reached at a lower PVC-E content: 30% for NBR-18 and NBR-26 against 20% for NBR-40. In all the composition ranges, the strain at break is much higher for vulcanizates based on the rubbers of higher polarity (NBR-26, NBR-40) than for the system based on NBR-18. This is evidence of a stronger contact between the blend components in the former two systems as compared to the latter one.

A difference between the structures of vulcanizates is well manifested in the plots of O300 versus composition. In the NBR-18 based composition, the a_{300} value monotonically increases with the PVC-E content, while the plots of the NBR-26-PVC-E and NBR-40-PVC-E systems have an S-like shape. In the region of a 30% polymer content, the modulus increases 6–7 times as compared to the values typical of the pure rubber and reaches a level characteristic of the PVC-E component. In the NBR-18 based system, the isolated coarse PVC-E particles play the role of a reinforcing filler. Compatible blends based on NBR-26 and NBR-40 exhibit phase inversion at a PVC-E content of 20–30% [26,29,30]. This implies the formation of a strong spatial framework of polymer molecules, which imparts high strength and hardness to the vulcanizates. Below this threshold concentration, the PVC particles are isolated from one another and produce no reinforcing effect. The size of the thermoplast particles is apparently comparable to the radius of curvature (below 10 pm) at the crack vertex, at which the reinforcing effect vanishes [26]. We may suggest that PVC is partly dispersed on a molecuar level, while the main part enters into molecular agglomerates not possessing the properties of a microphase. These systems are characterized by the especially large role of the transition layer between phases.

The pattern of variation of the mechanical properties of blends with the composition shows correlation with the behavior of the ozone resistance of vulcanizates (Fig. 3) tested at a strain of 30% and a temperature of 60°C (for the NBR-PVC-E blends heat-treated at 120°C). These conditions of testing were selected because the 30% tensile strain is close to the critical level featuring the maximum rate of the ozone-induced rubber degradation.

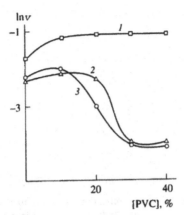

FIGURE 7.3   The plots of stress relaxation rate in ozone vs. PVC-E content in vulcanized blends based on (1) NBR-18, (2) NBR-26, and (3) NBR-40 (T = 333 K; ε= 30%).

A comparison of the pattern of variation of the σ_{300} stress (Fig. 7.2) to the curve of the stress relaxation rate versus composition (Fig. 7.3), which also has an S-like shape, shows that the formation of a continuous spatial framework of PVC molecules in the compatible systems leads to a jump like increase in the ozone resistance by two orders of magnitude: the rate of stress relaxation due to the ozone-induced cracking decreases to a level corresponding to the physical stress relaxation rate.

The limited-compatibility system NBR-18-PVC-E exhibits, irrespective of the composition, a higher stress relaxation rate in the ozone-containing atmosphere (i.e., a lower ozone resistance) as compared to that of the pure NBR-18 rubber. Apparently, the isolated coarse PVC particles may serve as stress concentrators in the extended elastomer matrix weakly bound to these particles via the transition layer. This circumstance favors the accelerated reaction between ozone and the double bonds of rubber [20].

Thus, our results obtained by the method of stress relaxation confirm the previous data that PVC provides protection of the compatible rubbers from the ozone-induced cracking only when its concentration in a rubber-PVC blend is not less than 30% [1–3, 6, 7]. Another condition for the protective effect of PVC consists in homogeneity of the system.

The replacement of PVC-E by PVC-S results in sharply increasing heterogeneity of the blend structure and decreasing ozone resistance of the vulcanized samples upon the treatment at 120°C (series B). Figure 7.4 shows the plots of stress relaxation rate in the ozone-containing

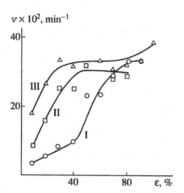

FIGURE 7.4   The plots of stress relaxation rate at 30°C in an ozone-containing atmosphere vs. strain for vulcanized 70: 30 blends of various NBRs (series I—III) with PVC obtained by low-temperature blending followed by the heat-treatment in a thermostat at 120°C (series B).

atmosphere versus deformation at 30°C. A sharp drop of the ozone resistance is evidence of a reduced interphase adhesion and easy phase separation. It should be noted that vulcanized samples prepared from the Paracril BJ robber exhibit a higher ozone resistance as compared to that of the domestic vulcanizates. Indeed, the former material is characterized by a low relaxation rate in the ozone-containing medium up to a deformation level of 40%, whereas the latter samples cannot withstand even smallest tensile deformations.

Vulcanizates possessing very high ozone resistance were obtained by the high-temperature blending of NBR containing 26–28% of AN units with PVC-S (series C). The temperature dependence of the stress relaxation rate in these samples measured in an ozone-containing atmosphere (Fig. 7.5) exhibit a degenerate ψ-transition except the blend with SKN-26 M (cf. with a curve for SKN-26 blended with PVC-E at low temperature and treated at 120°C in series A). Degeneracy of the ψ-transition can be considered as evidence of a higher degree of homogeneity provided by the high-temperature blending. Apparently, PVC in the samples of series C is dispersed either on a molecular level or in the form of very small molecular associates. The transition layer has a small total volume and its properties are not significantly manifested in macroscopic characteristics such as the stress relaxation rate. The high ozone resistance of these blend

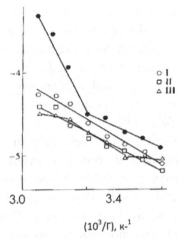

$$(10^3/\Gamma), \text{к-}^1$$

FIGURE 7.5 Temperature dependence of the rate of stress relaxation in an ozone-containing medium for 70: 30 blends of various NBRs (series I—III) with PVC obtained by high-temperature blending for 8 min at 170°C (Series C, e = 30%; see the text for explanations).

systems is confirmed by data on the effect of deformation on the rates of the physical relaxation and the relaxation process in ozone (Fig. 7.6). As seen, the stress relaxation in both media proceeds apparently by the same mechanism.

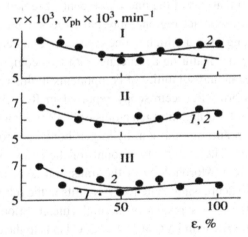

FIGURE 7.6 The plots of (*1*) the physical relaxation rate and (2) the stress relaxation rate in an ozone-containing atmosphere at 30°C versus strain for vulcanized 70: 30 blends of various NBRs (series I—III) with PVC-C obtained by high-temperature blending at 170°C (series C).

The mechanism of the ozone-protective effect of PVC in a highly homogeneous system is probably somewhat different from that described above. For a nearly molecular mixing, the homogeneity of a thermoplast distribution in an elastomer matrix markedly increases. The matrix has proved to be rather uniformly filled with stretches of PVC macromolecules, which hinder the opening of submicrocracks and inhibits the ozone penetration deep into the sample.

A difference in the structure of vulcanizates in series B and C is also manifested by their mechanical properties. Insufficiently homogenized samples of series B, having poorly developed loose transition layer, are characterized by low ultimate strength (not exceeding 7.7 MPa), whereas the strength of homogeneous samples in series C is 2.5–7 times higher (reaching 17–24 MPa). The most ozone-resistant samples in series B (Fig. 7.4) exhibit a maximum strength of 7.7, 2.6, and 3.6 MPa in series I-Ш, respectively. In heterogeneous blends, both characteristics are apparently related to the properties (strength) of the transition layer.

The replacement of SKN-26 by SKN-18 (incompatible with PVC) in series C leads to a dramatic decrease in the ozone resistance of vulcanizates (the samples are broken within 3–5 min). Therefore, even forced high-temperature blending of thermodynamically incompatible components cannot ensure the obtaining of ozone-resistant blends. This is related to morphological features of the interphase contact regions.

The interphase contact regions may have the structure of two types [30]. We may suggest that a high ozone resistance is inherent in the elastomer-polymer pairs featuring a structure with two-component transition layer and the segmental solubility of components in this region. The data of Fourier-transform IR spectroscopy reported in Ref. [9] showed that chlorine atoms of PVC and nitrogen atoms of NBR participate in a dipole-dipole interaction and, in addition, hydrogen bonds are formed between the two polymers. These contacts account for the compatibility of polymers in this system. Morphology of the transition layer may apparently sharply vary with decreasing content of AN units in the rubber component.

Heating of the rubber results in thermal vulcanization [31, 32]. The plasticization of NBR with PVC at 170–175°C leads to the chemical interaction between the active groups of rubber and thermoplast, which results in the formation of grafted copolymers [33–35].

The above notions concerning the structure of NBR-PVC blends of various homogeneity are also confirmed by the results of our investigation of the rotational mobility of nitroxyl radicals 1 and 2 as function of the temperature (Fig. 7.7). As seen, the rotation correlation times τ_c of both radicals are smaller in the blends with SKN-18 than with SKN-26, which is explained by the matrix being more rigid in the latter case.

The plots of τ_c versus temperature obtained for the TEMPO radical (Fig. 7.7) exhibit breaks corresponding to the relaxation transitions in the interphase layer (30–35°C) and in NBR (50–60°C) [27, 28] and large elements of the spatial network of PVC (~80°C). Therefore, the TEMPO radical adsorbs in a sufficiently large amount within the most mobile regions of these structural elements. The positions of these breaks allow us to judge on the blend structure and the size of homogeneous regions in the polar component framework. The blends of SKN-18 obtained by any method are highly heterogeneous, whereby PVC occurs in the form of both small and coarse inclusions, possessing the properties of macrophase,

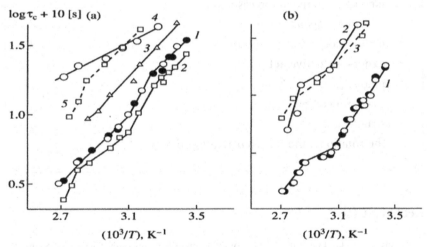

FIGURE 7.7 Temperature dependence of the rotation correlation time of nitroxyl radicals 1 and 2 in vulcanized 70:30 NBR- PVC blends: (a) SKN-26 with (*1*) radical 1 and PVC-E (open symbols) or PVC-S (60 rpm, black symbols), (*2*) radical 1 and PVC-S (30 rpm), (*3*) radical 2 and PVC-E, (*4, 5*) radical 2 and PVC-S (30 and 60 rpm, respectively); (b) SKN-18 with (*1*) radical 1 and PVC-E (open symbols) or PVC-S (60 rpm, black symbols), (*2*) radical 2 and PVC-E, (*3*) radical 2 and PVC-S (60 rpm).

as indicated by the transition in the glass transition region of bulk PVC at 80°C. Note that the same transition is observed in a heterogeneous blend of SKN-26 with PVC obtained upon the high-temperature blending at a reduced mixer speed (30 rpm, see sample 7). In this blends is observe the transition in the region of 60°C – is λ-relaxations related to formation of fluctuation structures involving *trans*-butadiene units [27, 28]. This is evidence of a high polydispersity and the presence of coarse particles of the thermoplast. The resulting vulcanizates exhibit very low ozone resistance. In contrast, the highly homogeneous ozone-resistant blends exhibit no breaks in this temperature region, which can be interpreted as indication of a sufficiently high degree of PVC dispersion in these systems.

KEYWORDS

- acrylonitrile-butadiene rubbers
- compatibility
- electron paramagnetic resonance
- interphase layers
- ozone-induced degradation
- ozone-protective action
- phase
- stress relaxation
- structure
- the rotational mobility (correlation time τ_c)

REFERENCES

1. Shvarts, A. G., Dinzburg, B. N., Combination of Rubbers with Plastics and Synthetic Resins, Moscow: Chemistry, 1972.
2. Mamedov, Sh.M., Yadreev, F. I., Rivin, E. M., Butadiene-Nitrile Rubbers and Related Elastomers, Baku: Elm, 1991.
3. Jorgensen, A. H., Frezer, D. Y., *Rubber World*, 1968, vol. 157, no. 6, p. 57.
4. Livanova N. M., Lyakin Yu. I., Popov A. A., Shershnev V. A., *Polymer Science, Ser. A* 49, 63 (2007) (Russian Journal) [*High molecular compounds, Ser.* A 49, 79 (2007)].

5. Livanova N. M., Lyakin Yu. I., Popov A. A., Shershnev V. A., *Polymer Science, Ser.* A 49, 300 (2007) (Russian Journal) [*High molecular compounds, Ser.* A 49, 465 (2007)].

6. Zateev, V. S., *Abstract of Cand. Sci. (Tech.) Dissertation,* Volgograd: Volgograd Polytechnical Inst., 1972.

7. Khanin, S. E., *Abstract of Cand. Sci.* (Chem.) Dissertation, Moscow, NHShP, 1984.

8. Morrel, S., *Rubber J.,* 1972, vol. 154, no. 12, p. 1956.

9. Zheng Xiaojiang, Pu Henry, H., Yang Yanheng, and Ziu Junfeng, *J. Polym. Sci., Part C: Polym. Lett.,* 1989, vol. 27, no. 7, p. 223.

10. Oganesov, Yu.G., Osipchik, VS., Mindiyarov, Kh.G., Raevskii, V. G., Voyutskii, S. S., *High molecular compounds,* Ser. A, 1969, vol. 11, no. 4, p. 896 (Russian Journal).

11. Oganesov, Yu.G., Kuleznev, V. N., Voyutskii, S. S., *High molecular compounds,* Ser. B, 1970, vol. 12, no. 9, p. 691 (Russian Journal).

12. Aivazov, A. B., Mindiyarov, Kh.G., Zelenev, Yu.V., Oganesov, Yu.G., Raevskii, V. G., *High molecular compounds,* Ser. B, 1970, vol. 12, no. 1, p. 10 (Russian Journal).

13. Kiseleva, R. S., Mindiyarov, Kh.G., Ionkin, V. S., Gubanov, E. F., Ushakova, G. G., Golikova, F. A., Zelenev, Yu.V., Voskresenskii, V. A., *High molecular compounds,* Ser. A, 1972, vol. 14, no. 9, p. 2078 (Russian Journal).

14. Tkhakakhov, R. B., Aivazov, A. B., Dinzburg, B. N., Zelenev, Yu.V., *High molecular compounds,* Ser. B, 1980, vol. 22, no. 11, p. 843 (Russian Journal).

15. Perepechko, I. I., Trepelkova, L. I., Bodrova, L. A., Bunina, L. O., *High molecular compounds,* Ser. B, 1968, vol. 10, no. 7, p. 507 (Russian Journal).

16. Kuleznev, V. N., Dogadkin, B. A., Colloid Journal, 1962, vol. 24, no. 5, p. 632 (Russian Journal).

17. Krisyuk, B. E., Popov, A. A., Livanova, N. M., Farmakovskaya, M. P., *Polymer Science, Ser.* A, 1999, vol. 41, no. 1, p. 94. (Russian Journal)

18. Popov, A. A., Parfenov, V. M., Krasheninnikova, G. A., Zaikov, G. E., *High molecular compounds, Ser.* A, 1988, vol. 30, no. 3, p. 656. (Russian Journal)

19. Zuev, Yu.S., Polymer Degradation under the Action of Aggressive Media, Moscow: Chemistry, 1972.

20. Popov, A. A., Rapoport, P. Ya., Zaikov, G. E., Oxidation of Oriented and Strained Polymers, Moscow: Chemistry, 1987.

21. Antsiferova, L. I., Valova, E. V., Chemical physics, 1994, vol. 13, no. 6, p. 89. (Russian Journal)

22. Antsiferova, L. I., Valova, E. V., *Polymer Science, Ser. A,* 1996, vol. 38, no. 11, p. 1215. (Russian Journal)

23. Synthesis and Properties of PVC, Zil'berman, E. N., Ed., Moscow: Chemistry, 1968.

24. Shtarkman, B. P., Plasticization of PVC, Moscow: Chemistry, 1975.

25. Popov, A. A., Livanova, N. M., Bogaevskaya, T. A., Farmakovskaya, M.P, RF Patent no. 95109654/09, 1996.

26. Kuleznev, V. N., Polymer Blends: Structure and Properties, Moscow: Chemistry, 1980.

27. Livanova N. M., Karpova S. G., Popov A. A., *Polymer Science, Ser.* A 51, 979 (2009) (Russian Journal) [*High molecular compounds, Ser.* A 51, 1602 (2009)].

28. Livanova, N. M., Karpova S. G., Popov A. A., *Polymer Science, Ser.* A 53, 1128 (2011) (Russian Journal) [*High molecular compounds, Ser.* A 53, 2043 (2011)].

29. Manson, J. A., Sperling, L. H., *Polymer Blends and Composites,* New York: Plenum, 1976.

30. Physical Chemistry of Multicomponent Polymeric Systems, Lipatova, Yu.S., Ed., Scientific Thought, Kiev, 1986, vol. 2 [in Russian].
31. Moiseev, V. V., Esina, T. I., Aging and Stabilization of Butadiene-Acrylonitrile Rubbers, Moscow, TsNIITE Neftekhim, 1978.
32. Devirts, E.Ya., Butadiene-Acrylonitrile Rubbers. Properties and Applications, Moscow: TsNIITE Neftekhim, 1972.
33. Berlin, A. A., Ganina, V. I., Kargin, V. A., Kronman, A. T., Yanovskii, D. M., *High molecular compounds,* 1964, vol. 6, no. 9, p. 1688 (Russian Journal).
34. Berlin, A. A., Kronman, A. G., Yanovskii, D. M., Kargin, V. A., *High molecular compounds,* 1964, vol. 6, no. 9, p. 1688 (Russian Journal).
35. Kronman, A. G., Kargin, V. A., *High molecular compounds,* 1966, vol. 8, no. 10, p. 1703 (Russian Journal).

CHAPTER 8

MORPHOLOGY AND PHYSICAL-MECHANICAL PARAMETERS OF POLY(3-HYDROXYBUTYRATE) – POLYVINYL ALCOHOL BLENDS

A. A. OL'KHOV,[1] A. L. IORDANSKII,[2] and G. E. ZAIKOV[3]

[1]*Plekhanov Russian University of Economics, Stremyanny per. 36, Moscow117997 Russia, E-mail: aolkhov72@yandex.ru*

[2]*Semenov Institute of Chemical Physics, Russian Academy of Sciences, Kosygin str. 4, Moscow, 119991 Russia*

[3]*Emanuel Institute of Biochemical Physics, Russian Academy of Sciences, Kosygin str. 4, Moscow, 119334 Russia*

CONTENTS

ABSTRACT

The structure of extruded films based on blends of polyvinyl alcohol and poly(3-hydroxybutyrate) (PHB) was studied for various compositions. The methods of DSC and X-ray analysis were used. As the phase-sensitive characteristics of the composite films, diffusion and water vapor permeability were also investigated. Processes linkages of water and swelling cause the first areas, the second – by processes of a relaxation and transition of structure of composites to a equilibrium condition. In addition, the tensile modulus and relative elongation-at-break were measured. Changes in the glass transition temperature of the blends and constant melting points of the components show their partial compatibility in intercrystallite regions. At a content of PHB in the composite films equal to 20–30 wt %, their mechanical characteristics and water diffusion coefficients are dramatically changed. This fact, along with the analysis of the X-ray diffractograms, indicates a phase inversion in the above narrow concentration interval. The complex pattern of the kinetic curves of water vapor permeability is likely to be related to additional crystallization, which is induced in the composite films in the presence of water.

8.1   INTRODUCTION

To vary the physicochemical, mechanical, and diffusion characteristics of the polyvinyl alcohol (PVA) and to widen the area of practical application [1–3], mixed compositions with a moderately hydrophilic polymer were proposed [4].

To improve the mechanical behavior of PHB and simultaneously to depress an expenditures in its production, the modification can be made through the PHB blending with other relevant polymers. Resulting polymer blends are potentially able to gain properties to be different from properties of parent blend-forming polymers.

Poly(3-hydroxybutyrate) (PHB) was used as the modifying polymeric component. The selection of this polymer is due to its biocompatibility with animal tissues and blood. Taking into account similar properties of PVA, one may expect that a new class of polymer materials for medical purposes will be created [5, 6].

A widespread procedure for regulating the drug release rate involves controlled changes in the balance of hydrophilic interactions in the polymer matrix at the molecular level. Therefore, regulation of structural organization at the molecular and supramolecular levels makes it possible to control the rate of drug deliver and hence to improve the therapeutic efficacy of new medicines.

8.2 EXPERIMENTAL

We studied the PVA (trade mark 8/27, Russia) containing 27% VAc; M = 3.8×10^4. The corresponding DSC curve shows two melting peaks at 130 and 170°C. As the other component, we used powdered PHB (Lot M-0997, Biomer, Germany) with M = 3.4×10^5; its melting point is 176°C, and the degree of crystallinity is equal to 69% (X-ray) or 78% (DSC).

The mechanical mixtures were prepared in the following proportions PVA:PHB = 100:0, 90:10, 80:20, 70:30, 50:50, and 0:100 wt %. The as-prepared mixtures were processed into films with a thickness of 60 ± 5 pm using an ARP-20 single-screw extruder (Russia) with a screw diameter of 20 mm and a diameter-to-length ratio of 25. In the extruder zones, the temperature was varied from 150 to 190°C.

The structure of the as-prepared films was characterized by DSC methods on a Mettler TA-4000 indium-calibrated thermal analyzer; the scanning rate was 20 K/min. The structure of the test samples was also characterized by X-ray analysis on an automated X-ray diffractometer (CuK$_a$-radiation, λ = 0.154 nm) with a linear coordinate detector (Joint Institute for Nuclear Research, Chemogolovka) [7].

The tensile modulus and relative elongation-at-break of the composite films were estimated using a ZE-40 tensile machine (Germany) at a cross-head speed of 100 mm/min; the gage width of the test samples was 10 mm. The water vapor permeability of the composite films was measured according to the standard beaker method [8] at 23 ± 1°C and a saturated water vapor pressure of 21 mmHg; the results were taken from five parallel experiments. The accuracy of weighing was equal to ± 0.001 g.

8.3 RESULTS AND DISCUSSION

The results of DSC studies of the composite films based on the PVA copolymer and PHB are summarized for various film compositions in Fig. 8.1

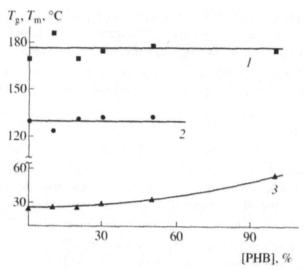

FIGURE 8.1 (*1*) High-temperature and (*2*) low-temperature melting points and (*3*) glass transition temperature of the composite films vs. concentration of PHB.

and Table 8.1. As is seen, the average glass transition temperature T_g of the blends lies between 24.1 and 53.9°C, the values corresponding to the glass transition temperatures of the starting PHB and PVA. This fact indicates a mixing of segments of these macromolecules in amorphous regions. The changes in T_g of the composites with respect to T_g of the starting polymers clearly indicate the compatibility of the components. However, the presence of a crystalline phase in both components at various content ratios suggests a limited interaction between the polymers.

TABLE 8.1 Characteristics of the Composite Films Based on PVA and PHB

PVA:PHB, %	Tm, °C	Pw* × 10⁸, [g·sm/ sm²·h·mmHg]	Cw** × 10³, [g/ sm³·mmHg]
100:0	129.3/170.0	280.0	37.0
90: 10	123.2/186.1	75.0	9.0
80:20	130.4/170.0	105.0	3.0
70: 30	132.0/175.0	115.0	1.0
50:50	132.0/178.0	125.0	1.4
0:100	176.0	0.33	0.1

Note: Two melting temperatures (T_m) of the composite films correspond to the two peaks in the DSC healing scans. *Water vapor permeability coefficient. **Water solubility coefficient.

Analysis of the nature of the double peak in the DSC melting curve of the PVA samples is beyond the scope of this work. However, one may attempt to explain this fact on the basis of published data [9–11].

This behavior may be related to a wide size distribution of crystallites, as well as to their metastable character. As was asserted in Ref. [10], the typical DSC curves of crystalline polymers show several melting peaks; their height and position on the temperature axis are controlled by the temperature-time conditions of the sample processing in the extruder.

The presence of the double peak in the DSC curves of the PVA films may also be explained by the effect of extrusion-induced orientation, which may lead to the development of various morphological structures with different temperatures of phase transitions. As was stated in Ref. [11], polymers may have various crystalline modifications with identical crystallographic parameters of the unit cell but different energy levels.

As was found, the positions of the high-temperature and low-temperature peaks in the DSC curves, which correspond to the melting points of the starting components, are almost independent of the blend composition and remain invariable in the whole concentration range under study (Fig. 8.1). However, the transient region characterizing the glass transition temperature of the PVA – PHB system assumes different positions on the temperature axis depending on the concentration of PHB. This situation is vividly illustrated in Fig. 8.1, where T_g of the blend is seen to increase with the content of PHB.

To reveal the morphology of crystalline regions in the composite films, the DSC measurements were combined with an X-ray study. As was shown for the entire composition range under study, the quasi-crystalline phase of the PVA copolymer is characterized by a well-pronounced orientation of the macromolecular axes along the texture axis, which coincides with the extrusion direction: the (110) X-ray reflection is located on the equator (Fig. 8.2a). In the crystal lattice of the PVA copolymer, the normal to the (110) plane is perpendicular to the c axis, along which the axes of macromolecules are oriented. In samples containing 10 and 20% PHB, the crystalline phase of PHB is almost fully oriented. In this case, the (020) reflection is located on the equator. The normal to the (020) plane is the b axis, and the a axis coincides with the direction of the him extrusion. The c axis of the unit cell of PHB, which is parallel to the directions of macromolecules, is perpendicular to the extrusion direction.

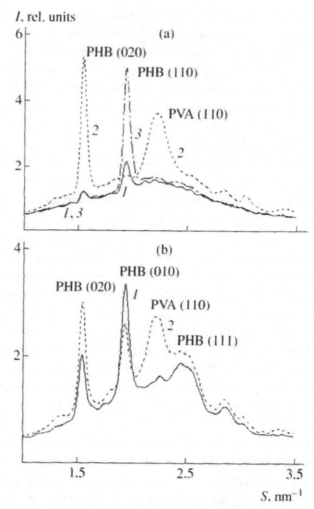

FIGURE 8.2   X-ray diffractograms of the films with a composition of (a) 80: 20 and (b) 70:30 wt % recorded (*1*) alone the orientation axis and (*2, 3*) at an angle of (*2*) 90° or (*3*) 20° to the orientation axis. S = 2 sin θ/λ, where θ is the X-ray scattering angle and λ is the wavelength.

X-ray diffractograms of films containing 30% and more PHB show the presence of the isotropic phase of PHB in appreciable amounts (Fig. 8.2b). The parameters of the unit crystal cells of the polymers in the blends were estimated. The crystalline phase of PHB is shown to be characterized by an orthorhombic unit cell: a = 0.576, b = 1.32, and c = 0.596 nm [10]. The crystalline phase of the PVA in the quasi-crystalline modification includes regions with a close packing of parallel chains (the γ modification); the parameters of the unit cell are a = 0.78, b = 0.253, and c = 0.549 nm [12]. In the corresponding X-ray diffractograms, these regions are associated with an X-ray reflection at S = 2.22 nm^{-1}.

The data of X-ray analysis allow one to conclude that the crystal lattice parameters of the components of the composite films are invariable; only the content of the isotropic component of the crystalline PHB phase is changed in the region of the phase transition.

The results of mechanical tests of the PVA-PHB composites (uniaxial tension) fully support the above conclusion that phase inversion takes place in the films containing 20–30% PHB. This reasoning is proved by an abrupt inflection in the dependences of the tensile strength σ_t (Fig. 8.3, curve 1) and elongation-at-break ε_t (Fig. 8.3, curve 2) on the composition of the blends, as well as by the presence of a minimum in the dependence of the tensile modulus E_t (Fig. 8.4) on the composition of the composite films.

According to the data presented in Figs. 8.3 and 8.4, the properties of the composite films are controlled by the PVA matrix when the concentration

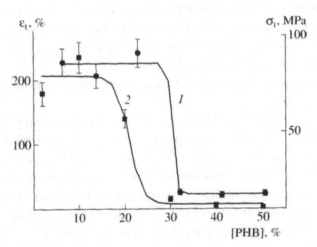

FIGURE 8.3   (*1*) Tensile strength σ_t and (*2*) relative elongation-at-break ε_t of the composite films vs. content of PHB.

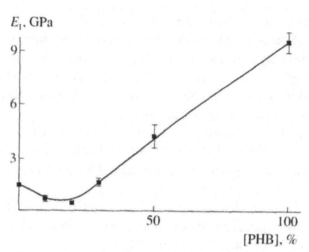

FIGURE 8.4   Tensile modulus E_t of the composite films vs. content of PHB.

of PHB is below 20% but determined by the PHB matrix at a concentration of 30% PHB and higher.

Almost invariable σ_t and ε_t values of the composite films before and after the phase inversion region may be explained as follows: particles of the dispersed phase do not serve as stress-concentrating sites in the matrix. This situation is possible when the dispersed phase is uniformly distributed within the matrix and the dimensions of its particles are rather small. This is usually observed in the case of a complete or partial miscibility of the polymer components [13, 14].

Let us consider the process of water vapor transfer through the PVA-PHB composite films.

Figure 8.5 presents the kinetic curves of vapor permeability for various PHB contents. One may distinguish three characteristic portions in these curves. The initial portion refers to a non-steady-state transport mechanism (Fig. 8.5b). In this region, the diffusion flow depends on time because the diffusion is related to the physicochemical process of the binding of water molecules on functional groups of PHB, which show a marked affinity to water (carbonyl groups [14], acetate and hydroxyl groups of the PVA [15, 16]).

The next (middle) portion in the kinetic permeability curves corresponds to quasi-steady-state transport, where the segmental mobility of

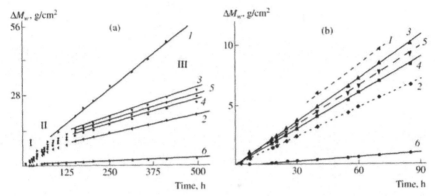

FIGURE 8.5   Kinetic curves of water vapor permeability per unit area of the composite films with the following compositions: (*1*) 100:0, (*2*) 90:10, (*3*) 70:30, (*4*) 80:20, (*5*) 50:50, and (*6*) 0:100. The Roman numerals refer to the three characteristic regions discussed in the text, (a) General pattern and (b) initial fragment of the kinetic curves.

PVA increases under the action of the diffusing water and an additional crystallization of the hydrophilic component is likely to occur [17]. Such phenomena were described in detail and analyzed in Ref. [18]. As the degree of crystallinity in the composite films increases, the rate of the effective flow of the diffusing component must decrease, and this trend is well seen in the last portion of the kinetic curves (Fig. 8.5a). Since newly formed crystallites are impermeable to water and therefore create an additional diffusion resistance, the slope of the last portion of the kinetic curves markedly decreases compared to the slope of the curves in the preceding portion.

The dependence of diffusion coefficients on the composition of the polymer blend (Fig. 8.6) shows a characteristic inflection at a content of 30% PHB, which corresponds to the phase inversion of the polymer matrix. In this region, one may observe an abrupt decrease in the tensile strength and relative elongation-at-break (Fig. 8.3). Strictly speaking, the traditional interpretation of diffusion equations in this region seems to be poorly justified and requires a detailed analysis, including consideration of the convective transfer mechanism [19]. The correlation between the transport and mechanical characteristics of the films indicates the important role of structural and morphological elements of the PVA-PHB mixed compositions.

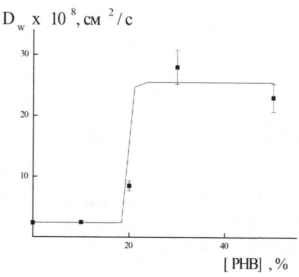

$$D_w \times 10^{\,8}, \text{см}^{\,2}/c$$

[PHB] , %

FIGURE 8.6   Water vapor diffusion coefficients D_w of the composite films vs. content of PHB.

The Table 8.1 lists the vapor permeability and solubility coefficients for the initial polymers (PVA and PHB) and composite films of various compositions estimated by the Daynes-Barrer method [20]. As follows from the Table 8.1, the solubility coefficient C" monotonically decreases with the increasing concentration of PHB. This result seems to be quite evident, since the amount of sorbed water depends on the nature and concentration of functional groups in the polymer.

The coefficient of water vapor permeability slightly depends on the composition of the blends. Since this parameter is controlled by diffusion, equilibrium water sorption, and structure of the films at the crystal level, the interpretation of the results is ambiguous. For example, the above decrease in the solubility may be compensated by a structural amorphization of the components of the composite films near the phase inversion point.

As the content of the PVA copolymer in the polymer blend is increased, the concentration of hydroxyl groups evidently rises. The hydrophilization of the PHB matrix (with increasing content of the PVA component) provides a monotonic increase in the water solubility coefficient without any visible inflection points and extrema (see Table 8.1). Even though the films may experience various structural rearrangements at the crystal level, as is evidenced by the DSC data, X-ray analysis, and mechanical tests, the water solubility in the PVA-PHB composite films is still sensitive only to changes in the ratio between the contents of the hydrophilic and moderately hydrophilic components. Therefore, the content of dissolved water is controlled by the nature and concentration of functional groups in the polymer blend.

As the content of PHB in the blends is increased, the concentration of hydroxyl groups decreases; that is, at the molecular level, hydroxyl groups with their high group contribution (according to van Krevelen) are substituted by ester groups of PHB with a much lower group contribution [21].

8.4 CONCLUSIONS

A blend design on the base of polymer components with essentially different hydrophilicities is of academic and industrial interest, primarily, for the control of such critical but scantily known characteristics as water permeability, its equilibrium sorption and diffusional mobility.

In spite of limited concentration interval of partly miscible blending (no more than 30%wt. of PHB) these PHB-PVA blends are of great interest as novel biodegradable films and coatings. Additionally, the blends based on water-soluble and biocompatible PVA and friendly environmental and biocompatible PHB could be treated as new generation of environmentally protected materials in packaging industry, agricultural application as well as biomedicine areas [22–24].

KEYWORDS

- blends
- diffusion
- DSC
- mechanical parameters
- poly(3-hydroxybutyrate)
- polyvinyl alcohol
- structure
- X-ray analysis

REFERENCES

1. Sudesh K., Abe H., Doi Y. Prog. Polym. Sci, 2000. vol. 25, no. 7, p. 1503.
2. Holmes, P. A., Developments in Crystalline Polymers II, Bassett, D. C., Ed., London: Elsevier Appl. Sci., 1988.
3. Barak P., Coqnet Y. Halbach T. R., Molina J. A. E. J. Environ. Quai, 1991, vol. 20, №. 1, p. 173.
4. Mergaert J., Webb A., Anderson C., Wouters. A., Swings J. Appl. Environ. Microbiol. 1993, vol. 93, no. 12, p. 3233.
5. Timmins M. R., Lenz R. W., Fuller R. C. Polymer, 1997, vol. 38, no. 2, p. 551.
6. Yoshie N., Azuma Y, Sakurai M., Ionoue Y. J. Appl. Polym. Sci. 1995, vol. 56, no. 1, p. 17.
7. Iordanskii A. L., Ol'khov A. A., Pankova Yu. N., Kosenko R. Yu., Zaikov G. E. Transport of water as structurally sensitive process characterizing morphology of biodegradable polymer system. in "Chemical reaction in condensed phase: the quantitative level". Ed. By Zaikov G. E. New-York. Nova Sci. Publ. 2006, pp. 139 – 151.
8. Godovskii Yu. K., Teplofizicheskie metody issledovaniya polimerov (Thermophysical Methods for Polymer Investigation) Moscow: Khimiya, 1976.
9. Polymer Blends. Paul, D. and Newman S. M. Eds. New York: Academic, 1979.
10. A. A. Ol'khov, A. L. Iordanskii, G. E. Zaikov Morphology and mechanical parameters of biocomposite based on LDPE – PHB. J Balk Tribol Assoc, 2014, 20(1), 101.
11. Iordanskii A. L., Rudakova T. E., Zaikov G. E. . Interaction of Polymers with Bioactive and Corrosive Media. Ser. New Concepts in Polymer Science. VSP Science Press. Utrecht – Tokyo Japan. 1994, 298p.
12. Iordanskii A. L., Olkhov A. A., Zaikov G. E., Shibryaeva L. S., Litvinov LA., Vlasov S. V. . Morphologicaly special features of poly 3-hydroxybutyrate/low density polyethylene blends. J. Polymer-Plastics Technology and Engineering, 2000, V.39, № 5. 783–792.
13. Iordanskii A. L., Olkhov A. A., Kamaev P. P., Wasserman A. M. Water transport phenomena in green and petrochemical polymers. Differences and similarities. Desalination, 1999, 126. 139–145.
14. Hassan CM. and Peppas N. A. Biopolymers. PVA Hydrogels. Advances in Polymer Science, Berlin: Springer-Verlag, 2000.
15. Rozenberg M. E. Polimery na osnove vinilatsetata (Vinyl Acetate-Based Polymers), Leningrad: Khimiya, 1983.
16. Water in Polymers, Rowland, S. P., Ed., Washington: Am. Chem. Soc.,1980.
17. Pankova Yu.N., Shchegolikhin A. N., Iordanskii A. L., Zhulkina A. L., Ol'khov A. A. and Zaikov G. E. The characterization of novel biodegradable blends based on polyhydroxybutyrate: The role of water transport. Journal of Molecular Liquids. 2010, v. 156, Issue 1, pp. 65–69.
18. Iordanskii A. L., Bonartseva G. A., Pankova Yu.N., Rogovina S.Z, Gumargalieva, K. Z., Zaikov G. E., and Berlin A. A. Current Status and Biomedical Application Spectrum of Poly(3-Hydroxybutyrate as a Bacterial Biodegradable Polymer. Current State-of-the-Art on Novel Materials. (Ed: Devrim Balköse, Daniel Horak, Ladislav Šoltés). V. 1. Ch. 12. 2014. Apple Academic Press. New York. P. 450.

19. Iordanskii A. L., Ol'khov A. A., Pankova Yu. N., Kosenko R.Yu., Zaikov G. E. Transport of water as structurally sensitive process characterizing morphology of biodegradable polymer system. in "Polymer and biopolymer analysis and characterization." Ed. G. E. Zaikov, A. Jimenez, New-York. Nova Sci. Publ. 2007, pp. 103–116.

20. Vogler E. A. Water in Biomaterials. Surface Sciences, Morra M., Ed., Chichester: Wiley. 2001.

21. Panov, A. A., Beloborodova, T. G., Anasova, T. A., Zaikov, G. E. Plant for Manufacturing Building Tile on the Basis of Polymeric Materials. J Balk Tribol Assoc 2008, 15(2), 243.

22. Bonartsev, A. P., Iordanskii, A. L., Bonartseva, G. A., Zaikov, G. E. Biodegradation and Medical Application of Microbial Poly(3-hydroxybutyrate). J Balk Tribol Assoc 2008, 15(3), 359.

23. Iordanskii, A. L., Chvalun, S. N., Shcherbina, M. A., Karpova, S. G., Lomakin, S. M., Shilikina, N. G., Rogozina, S. Z., Zaikov, G. E., Chen, X. J., Berlin, A. A. Impact of Water upon Structure and Segmental Mobility of the PHBV-SPEU Blends. J Balk Tribol Assoc 2013, 19(1), 144.

CHAPTER 9

PROSPECTIVE PRODUCTION WAYS OF NEW HEAT-RESISTING MATERIALS BASED ON POLYIMIDES

E. T. KRUT'KO[1] and N. R. PROKOPCHUK[2]

[1]Professor (BSTU); [2]Corresponding Member of Belarus NAS, Professor, Head of Department (BSTU), Belarusian State Technological University, Sverdlova Str.13a, Minsk, Republic of Belarus, E-mail: v.polonik@belstu.by

CONTENTS

ABSTRACT

This chapter is devoted to receiving and research of materials on the basis of industrially made different kind of polyimide, including chemical modifying by reactivity polyfunctional compounds. Existence in macromolecules of polyamide acids reactive carboxyl, amide and amino groups o, capable to interact with multifunctional monomeric and oligomeric modifiers, gives the chance of receiving the polyimide materials possessing properties of sewed polymers. As a result, provides improvement of strength properties and thermal characteristics of new polymeric materials for practical aimes.

9.1   INTRODUCTION

Entire prehistory of heat-resistant polymer investigations led to the believe that the most successful structure can be considered a rigid structure consisting of a benzene ring firmly connected to two five-membered nitrogen-containing ring. Polymers of this kind have one common name of aromatic polyimides (PI).

For the past 60 years the polymers of this class are the most commonly used heat-resistant polymers according to the complex of unique characteristics. They firmly hold the primacy among the materials being a source of a huge range of products for all branches of science and technology. The production of the tapes, fibers, coatings, paints, plastics, membranes, composites, bonding, foamed and other materials is based on polyimides. Each of them can work in extremely high thermic conditions [1–5].

9.2   EXPERIMENTAL

Polyimide films were the first commercial material used to create highly heat insulators. Currently polyimide films release in the world is at a level of more than 1000 tons per year. They are used as insulators of such electrical items as cables, generators, motors and other components and assembly parts used at elevated temperatures [6, 7].

High stability of the surface layers of polyimide films determines their preferential use for the manufacture of resistors. Variation on the surface resistance of polyimide films is 3 times lower than that of pyroceramic substrates.

Polyimide films are used in the manufacture of heating elements of devices, heat-resistant coatings printing plates, wires and cables. Protective film with adhesive coating protects conductive pattern of flexible printed circuits against corrosion preventing short circuits and accidental conductors from contact with metal surfaces of the equipment.

It should be noted that the need for film coatings in the microelectronics has dramatically increased in recent years. The special emphasis is given to the using of ultrathin films with low dielectric constant.

Some of the polyimide films are optically transparent in the visible and ultraviolet regions of the spectrum. It allows their using in optical communications technology.

Polyimides are used in the manufacture of elongate fiber optics and microfilters. A number of polyimide films and film coatings is used in various systems of liquid crystal displays.

Metallized polyimide films are of great importance in wireless communication systems as well as in the production of various resistors and capacitors. Their scope goes beyond electricity designs, and especially where it is necessary to use a durable, flexible and heat-resistant substrate for mirrors, screens and reflectors for energy flows of wide range. The polyimide film may be combined with a metal foil with an adhesive or a metal is sprayed directly onto the film surface. In some cases the metal is introduced into the polyimide film at the stage of prepolymer. For example certain amounts of gold or silver, are introduced as a fine dispersion in a solution of a polyamic acid. The film with a homogeneous distribution of the metal throughout the whole product volume is obtained after its imidization and its surface treatment [8, 9].

As for the surface metallization, the polyimide films are used in space technology, mirrored surfaces of solar panels and components of solar technological equipment, as well as the fabrication of multilayer protective fabrics for suits of astronauts, firefighters and emergency crews at hot shops and nuclear power plants.

Particular attention is being given to exploitation of ultrathin films of Langmuir – Blodgett (LB). Polyimide LB films are widely used in the construction of three-layer photodiodes, photolithographic purposes, in the process of stabilization of alternating voltage switching systems in the memory. It should be noted that the soluble polyimides are required as the photosensitive elements of both negative and positive types. This trend of the polyimide materials is considered to be one of the priorities. Nor can we fail to dwell on the polyimide fibers.

During the intensive search of new heat-resistant fiber-forming polymers in the former Soviet Union the greatest scientists in the field of polymer chemistry Academician V. V. Korshak, RAS Corresponding Member M. M. Coton, A. N. Pravednikovym and many others laid the theoretical basis for the creation of polyimide fibers which have been successfully implemented in industrial scale. Until now the heat-resistant fiber Arimid T remains the most heat-resistant synthetic fiber. The range of its working efficiency is from −27°C till +450°C. These fibers are nonflammable; retain elasticity and strength characteristics at liquid nitrogen temperature to the full extent. They are the only fibers that can withstand integrated doses of nuclear and ultraviolet irradiation to 3000 Mrad without significant change of operational properties. For polyimide fibers and fabrics almost complete recovery of the elastic deformation is characterized at elevated temperatures.

The combination of the unique properties of PI fibers allows their use for equipment operating for long periods at elevated levels of radiation and temperature; for the reinforcement of rubber products, GRP laminated structural materials; for the manufacture of uniforms for workers in the area of high radiation activity and high temperatures; for creation of space filters for cleaning of hot gases and corrosive liquids. Fibers obtained from optical fibers for fiber optics [10–15].

Belarus has being taken part in many polyimides synthesis researches, fibers and films derivation, study of their properties [3, 15–22]. Research results were a prerequisite for the development of new materials in microelectronics and their introduction into the production technology of large and very large scale integrated circuits at PD "integral" with a significant economic effect.

The polyimide foam manufacture is one of the priorities in the development of advanced materials. The unique qualities of polyimides have attracted attention of developers to create foam protective structures for high-speed vehicles, primarily for aircraft and spacecraft. The challenge of the protecting of ship's crew as well as scientific equipment at unmanned flight was developed long before the implementation of flight at speeds of the order of tens of thousands of kilometers per hour. The increased heat and flame resistance, low density of inner cladding cabins, flexibility and resilience of the various elements of isolation and the use of multilayer insulation in the seamless stitching of individual components are the main protection requirements. The foam insulation should not burn, give off toxic products of thermal degradation of the polymer and smoke in extreme emergency situations. The polyimides are the most suitable polymers suited these requirements. The various technological methods have been developed for their manufacture. Given the specific product characteristics polyimide foams may have a wide range of different properties (density, thermal, acoustic and fire- resistance, mechanical and material yield strength, environmental friendliness).

The polyimide foams are used in the production of insulation panels for covering booths of supersonic aircraft and manned spacecraft.

In unmanned aerial vehicles (satellites) thermal mode of polymer operation provides the range from −60 to +180°C, the mechanical load is 3 times more than the overload valid at start of manned spacecraft, and background radiation depends on the specific conditions of the satellite. The polyimide foams protection provides continuous television and radio communication, reliable storage of scientific information and coordinated work of all electrical, electronic and optical equipment of space laboratories.

Nowadays there are real predictions to create large structures of high strength and minimum weight by using reinforced hollow carbon fibers connected with ultrathin semiconductor hybrid polyimide films in space. These solutions allow you to create solar panels of huge capacity to serve the different spacecraft.

The polyimide foams are widely used in microelectronics as dielectrics with a very low dielectric constant, protective sensor coatings. stress buffers to compensate vibration load of many components of integrated circuits in extreme situations.

However, some additional problems required the further material's improvement arise in spite of the achievement of the basic parameters of suitability of these materials to meet the stringent requirements. The chemical modification of polyimides is one way of solving this problem.

Outstanding properties of polyimides allowed in recent years to create and implement a new generation of membranes which are used for the separation of gases, vapors and liquids. Polyimide membranes differ with extremely high resistance to almost all chemical agents. Their thermal stability allows to realize separation processes for a long period of time at high temperatures. The high selectivity is indispensable for gas separation. The ability to obtain highly selective and at the same time permeable to water and organic substances membranes is explained by special membrane molecular design of polyimides.

The implementation of this problem is possible on the macromolecular level when the methods of structures synthesis with well- articulated rigid portions of the polymer backbone which create a calibrated interchain packing with a very narrow distribution of free volume has been developed.

Fundamental theoretical development and extensive experience with numerous practical synthesis of polyimides polycondensation techniques [10–13] allowed to use in practice more than one hundred polyimides with diverse chemical structure to create membranes of diverse functionality. Polyimide membrane show the highest effect in the process of the separation and purification of gas mixtures (particularly with "simple gases" such as hydrogen, helium, carbon dioxide and some other petrochemical production gases).

Recent research in polyimide membrane showed that because of their high chemical inertness and stability they are promising even in the medical industry for creating artificial organs e.g. the membranes of fluorinated polyimides are tested in the apparatus of artificial light. This material for vascular oxygenation compared with silicone coatings shows a 4-fold improvement in oxygen and carbon dioxide gas exchange and the good compatibility with blood elements.

Currently, polyimide membranes industry leaders are Japan and the United States.

Nowadays up to 500 various composite materials and products of multiple types based on polyimide plastics are annually patented. The contained polyimides and their derivatives ranges from 5 to 100%.

For example strong and heat-resistant composite materials with operating temperatures up to 500°C have been created. Such composite imidoplastics are used in aerospace industry products (i.e., tips for nose cones and leading edges of the wings of supersonic aircraft, gas rudders, nozzle liners, flue missile components and aircraft engines.)

A brand AURUM polyimide has been developed in Japan. It has been assigned to the category of high performance plastics. It is stable with no weight loss up to 500°C, resistant to radiation and inert to nearly all reactants. Main applications of the polyimide composite material are structural elements and slip components.

The imidoplastics with carbon fillers are competitive in the manufacture of components and mechanisms in automotive engineering. They can be used as substitutes for aluminum and titanium alloys at high speeds, temperatures and loads, energy insulation for nuclear reactors [8].

The polyimide varnishes are successfully used for enameled wire of copper, aluminum, steel.

With regard to recent developments the emergence of new problems is associated with microelectronics [15]. This is explained by the fact that this area now requires highly soluble polyimide materials with hue, flexibility, high transparency of coatings, high adhesive characteristics of the sprayed metals layers, low shrinkage.

The researches in the synthesis, the relationship of the chemical structure, the structure and performance of new properties developed at the Department of THC and MRP film-forming polyimide compositions in order to obtain on their base some functional materials of different assignment for the microelectronics industry were dedicated to resolve these complex problems [16].

Three PhD (candidate) and one doctoral dissertation theses on this topic were defended at BSTU. More than 20 patents of the Republic of Belarus were received; more than 50 articles in scientific journals were published. The monograph has been published. A group of authors was awarded by the National Academy of Sciences of Belarus for the series of papers "Polyimides. Synthesis. Properties. Application."

It should be noted that polyimides synthesis requires a good theoretical training, highly qualified specialists in the field of the experimental chemistry to carry out fundamental researches that should precede the successful solution of specific practical problems.

The following achievements could be marked as an example of the new materials creation based on serially produced polypyromellitimides at the Department of THC and MRP of BSTU during last five years:

1. A new composite film material with improved strength and adhesion properties to a layer of such deposited metals as aluminum and copper was proposed on the basis of a detailed researches of UV, IR, EPR of laws spectroscopy of prepolymer polyimide complexation with metal-containing modifying components (ferrocenium hexafluorophosphate and cobalt acetylacetonate) [17–19].

2. A solution of creation of soluble polyimide systems by synthesizing fragmented polypyromellitimides with bulky and curved portions of the polymer chains reducing intermolecular interactions and their subsequent chemical cyclodehydration leading to the solubility of macromolecules was proposed to enable the application of polyimides in gallium and arsenide technology of microelectronic devices [20].

3. The calculations of geometrical parameters of polyimide macromolecules fragments were performed to explain the solubility reasons of bloksopolyimides. It was shown that such a structure of the polymer molecules consists of strongly folded conformation with high conformational parameter. It provides a more open structure of the polymer and facilitates its dissolution [21, 22].

4. Variety of requirements for the properties of polyimide materials is often achieved by using complex modifiers. This approach have been used to manufacture the low-shrinkage polyimide compositions for potting compounds and protective layers by co-administration of aerosil prepolymer, gadolinium oxide and crosslinking agent – bysmaleimide.

9.3 CONCLUSION

The manufactured polyimide film-forming materials have successfully passed the pilot-scale test at OJSC "Integral" and can be used in modern microelectronics.

Currently the department of the TNC and MRP at BSTU as well as BSU and BSUIR continue their researches in the field of polyimides.

KEYWORDS

- block co-polyimides
- geometrical parameters
- modification
- polyimide
- properties
- structural characteristics

REFERENCES

1. Polyimides is a class of heat-resistant polymers. M.I. Bessonov et al.; under Society. Ed. M.I. Bessonova. – Leningrad: Nauka, 1983. 308.
2. Buller, K.U. Heat and heat-resistant polymers. K.U. Buller. Moscow: Khimiya, 1984. 530.
3. Polyimides. Synthesis, properties, and application. E.T. Krut'ko et al.; under Society. Ed. N.R. Prokopchuk. Minsk: Belarusian State Technological University, 2002. 303.
4. Mihailin, A. Heat resistant polymers and polymeric materials. Y.A. Mihailin. St. Petersburg.: Profession, 2006. 623 p.
5. Synthesis and characterization of soluble polyimides from 1,1'-bis(4-aminophenyl)-cyclo-hexane derivatives. M. H. Yi et al. Macromolecules. 1997. Vol. 30, No. 2. 5606–5611.
6. Synthesis and properties of novel photosensitive polyimides containing chalcone moiety in the main chain. K. Feng et al. J. Polym. Sci. Part A, Polym. Chem. 1998. Vol. 36, No. 5. 685–693.
7. Coton, M. Role of aromatic polyimides in modern science and technology. M. Coton. Journal of Applied Chemistry. 1995. T. 68, no. 5. 882–826.
8. Sazanov, N. Applied value of polyimide. Sazanov N..Journal of Applied Chemistry. 2001. T. 74, no. 8. 1217–1233.
9. New fluorine-containing polyimide. A.L. Rusanov et al. Polym. comp. Ser. A. 2006. V. 48, № 8. 1527–1530.
10. New polyimides, carboxyl. A.L. Rusanov et al. Polym. comp. Ser. B. 2006. V. 48, № 5. 859–863.
11. Kuznetsov, A.A. Synthesis of polyimides in the melt of benzoic acid: Author. dis. Dr. Chem. Sciences: 02.00.06. A. Kuznetsov; Inst synthetic Polymeric Materials. N. Yenikolopov RAS. Moscow, 2008. 41.

12. Kuznetsov A., One-step high-temperature synthesis of polyimides in the melt of benzoic acid: the reaction kinetics modeling stage polycondensation and cyclization. A.A. Kuznetsov, A. Tsegelsky, P.V. Buzin.Polym. comp. Ser. A. 2007. T. 49, № 11. 1895–1904.

13. Reflecting conductive polyimide films prepared by chemical plating. S.K. Kudaykulova et al. Proceedings of the 17th Mendeleev Congress on General and Applied Chemistry, Kazan, 21–26 September. 2003: Abstracts: 3 t. Kazan State. Univ; Editorial Board.: N.P. Liakishev et al. Kazan, 2003. T. 3. 224.

14. Conductive film-forming compositions based on a mixture of polyaniline and polyimide. T.K. Myaleshka et al. Polym. comp. Ser. A. 2009. T. 51, № 3. 447–453.

15. Yakimtsova L.B. Synthesis of aromatic polyimides film forming a network structure: Author. dis. Cand. chem. Sciences: 02.00.06. L.B. Yakimtsova; Belarusian. Reg. Univ. Minsk, 1997. 17.

16. Prokopchuk N.R. Correlation of the circuit configuration of the macromolecular structure and thermal properties of oriented poliarilenimidov: dis. Cand. Physics and Mathematics. Sciences: 01.04.19. N.R. Prokopchuk. L., 1977. 160 p.

17. Hloba, N. Metalcontaining polyimides. N. Hloba, E. Krutko, S. Bogushevich.8th International Technical Symposium on Polyimides and High Performance Polymers STEPI 8, Montpellier, 9–11 June 2008. Un. Montpellier; ed. M. Aba-die et al. Montpellier, 2008. 291–300.

18. Electrorheological fluids based on the modified aromatic polyimides. N. I. Hloba et al. Electrorheological Fluides and Magnetorheological Suspensions: abstracts of 11th International Conference, Dresden, 12–16 September 2008. Institute of physics. Dresden, 2009. 18.

19. Electrorheological fluids based on the modified aromatic polyimides. A. Karabko et al. J. Appl. Phys. 2009. Vol. 148. 48–53.

20. Zhdanuk, E. N. Thermal imidization of polyimide compositions. E. N. Zhdanuk, N. I. Hloba, E.T. Krutko. E-MRS: abstracts of European Materials Research Society, Warsaw, 14–18 September 2009. Warsaw University of Technology. Warsaw, 2009. 215–216.

21. Soluble block copolyimides. A.I. Globa et al. Reports Nat. Acad. Sciences of Belarus. 2011. T. 55, № 3. 79–82.

22. Polyimide composition for protection of crystals of semiconductor devices and integrated circuits: a stalemate. 11322 Resp. Belarus, IPC (2006) C08 L 79. 00. E.T. Krut'ko, A.I. Globa, J.N. Galiyev, T.A. Zharskaya, I. Globa; Belarusian applicant. Reg. Tehnol. University, Nauch. Research. Rep. UNITA. pre "Minsk Research Institute of Radio." № A20070020; appl. 11.01.2007; publ. 21.08.2008. Afitsyyny bulletin. Nat. tsentr intelektual. ulasnastsi. 2008. № 6. 112.

CHAPTER 10

PROTECTIVE PROPERTIES OF MODIFIED EPOXY COATINGS

N. R. PROKOPCHUK,[1] M. V. ZHURAVLEVA,[2] N. P. IVANOVA,[3] T. A. ZHARSKAYA,[4] and E. T. KRUT'KO[5]

[1]Professor (BSTU), Corresponding Member of Belarus NAS; [2]PhD student (BSTU); [3]Assistant Professor (BSTU); [4]Assistant Professor (BSTU); [5]Professor (BSTU), Belarusian State Technological University, Sverdlova Str.13a, Minsk, Republic of Belarus

CONTENTS

ABSTRACT

This chapter is devoted to the development and research of new film materials based on epoxy resins with improved properties. Evaluation of protective properties is determined by electrochemical and physico-mechanical methods of research that provides the most complete picture of the corrosion processes occurring under the paint film. These studies allowed to adjust the coating composition to the individual applications in order to achieve a high degree of protection of metal surfaces.

10.1   INTRODUCTION

Epoxy materials have superior performance properties, so they have been used to produce high quality coatings. Coating materials based on epoxy oligomers are used to prepare responsible coatings for various purposes, chemical-resistant, water-resistant, heat-resistant and insulating coatings. They are characterized by high adhesion to metallic and nonmetallic surfaces, resistance to water, alkalis, acids, ionizing radiation, low porosity, small moisture absorbability and high dielectric properties [1]. However, there are a number of outstanding issues to improve the barrier properties of paints based on epoxy resins, which limit their wider use in aeronautical engineering as well as in engineering and shipbuilding [2]. Chemical structure of epoxy resins provides ample opportunities to control their properties by introducing modifying additives, to achieve maximum compliance with the requirements of the resulting material.

10.2   EXPERIMENTAL

The aim of this study was to develop and research new film materials based on epoxy diane resin with improved barrier properties.

The object of the study was commercially produced epoxy resin E−41 in solution (E−41 s) (TU 6−10−607−78), which is a solution of the resin E−41 with a mass fraction (66 ± 2) % in xylene (GOST 9410−78, GOST 9949−76) with acetone (GOST 2768−84) at a ratio of 4:3 by weight. Resin solution E−41 in a mixture of xylene and acetone (resin E−41 s) is used for the manufacture of paints for various purposes. Resin E−41 s refers to medium molecular weight (MW 900−2000) epoxy diane resin. Its density is 1.03−1.06 g/m^3. Product of copolycondensation of low molecular epoxy resin E−41 with difenilol propane is represented by the formula shown in Scheme 10.1

Physico-chemical characteristics of the resin E−41 s are shown in Table 10.1.

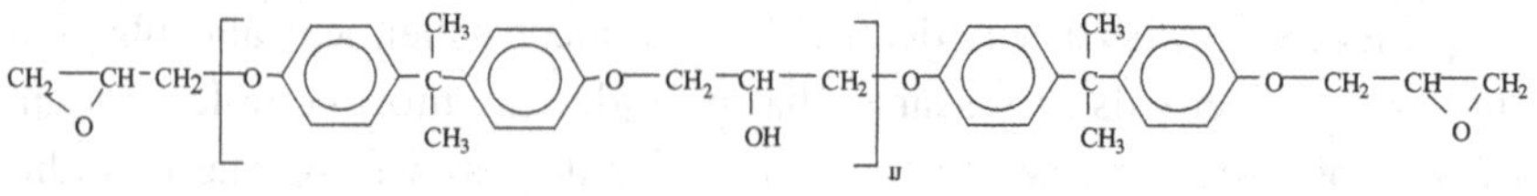

SCHEME 10.1

TABLE 10.1   Physico-Chemical Characteristics of the Resin E- 41 s

Index name	Norm on the highest quality category
Appearance of the resin solution	Homogeneous transparent liquid
Appearance of the film	Poured on glass is clean. Slight rash
Color on iodometric scale, mg I2/100 cm³, not darker than	30
Viscosity by viscometer VZ −246 (OT −4) with a nozzle diameter of 4 mm at a temperature of (20.0 ± 0.5)°C, c	80–130
Volume of solids, %	66 ± 2
Mass fraction of epoxy groups in terms of dry resin	6.8–8.3
Mass fraction of chloride ion (in terms of dry resin), %, not more than	0.0045
Mass fraction of saponifiable chlorine (in terms of dry resin), %, not more than	0.25

As a modifying component p-aminophenol (pAPh) was used. It is known that aminophenols are corrosion inhibitors. The presence of the aromatic ring and the amide and hydroxyl functional groups in the molecule of p-aminophenol dictates the fundamental possibility of using of this compound as a modifier of epoxy oligomer, allowing to increase the corrosion resistance of the coating.

The film-forming composition was prepared by introducing into E−41 s 10% solution of p-aminophenol in dimethylformamide at a concentration range of 0.5−5.0 wt. %, followed by stirring until smooth. Hardener of brand E- 45(TU 6−10−1429−79 with changing. № 2) − low molecular weight polyamide resin solution in xylene − in the amount of 14% by weight of resin solids was used. Films on various substrates were cast from the above solutions. Curing of modified epoxy diane compositions were carried out at a temperature of 110°C for 140 min.

Adhesive strength of the formed coatings were determined by the standard method in accordance with ISO 2409 and GOST 15140−78 by the cross-cut incision with the kickback. The essence of the method in the application of cross-cut incisions on the finished coatings with the tool

"Adhesion RN" and a visual assessment of the condition of the coatings after the impact action of the device "Kick–Tester" exerted on the opposite side of the plate at the site of incision. Coating condition is compared with standard classification and measured in points.

Impact resistance of coating samples were evaluated using the device "Kick–Tester" in accordance with ISO 6272 and GOST 4765–73.

Method for determining the impact strength of films (measured in centimeters) is based on the instantaneous deformation of the metal plate coated with the paint by the free fall of load on the sample and is realized through the device "Impact Tester" which is intended to control the impact strength of the polymer, powder and paint coatings.

The hardness of paint coatings was determined with a pendulum device (ISO 1522). The essence of the method consists in determining the decay time (number of oscillations) of the pendulum in its contact with the painted coating. Hardness is determined by the ratio of the number of oscillations of the coated sample to the number of oscillations of the sample without coating.

Flexural strength of the coatings was determined by a device SHG1 (ISO 1519 GOST 6806–73) by bending of the coated sample around test cylinders since larger diameters at an angle of 180°. On one of the cylinder diameters coating either cracks or breaks or peels. In this case, the paint has an elasticity of previous diameter test cylinder of the device in which it is not destroyed. Viewed count is in the radii of curvature in millimeters.

As can be seen from Table 10.2 at a curing temperature of 110°C of the obtained modified compositions with the modifier content of 0.5 to 2.0% hardness of the coating and adhesion is improving and impact strength is increasing.

Protective properties of coatings are determined by the sum of physical and chemical properties, which can be reduced to four basic characteristics:
- Electrochemical and insulation properties of coatings;
- The ability of films to retard diffusion and carrying corrosive agents to a metal surface;
- The ability of coatings containing a film forming, pigments or inhibi
- Adhesion and mechanical properties of the coatings.

TABLE 10.2 Adhesion and Strength Properties of Epoxydiane Coatings on Steel Substrates (Warming to 110°C, 140 min)

pAPh content, %	Hardness, rel. u.	Adhesion to Steel, points	Tensile impact, cm
0.0	0.66	1	2.5
0.5	0.69	0	5.0
1.0	0.73	0	10.5
2.0	0.70	0	7.5
3.0	0.67	1	3.0
4.0	0.63	2	6.0
5.0	0.47	2	1.0

All these properties are interrelated and influence each other mutually. Deterioration of properties of the film as a diffusion barrier will immediately lead to a reduction in adhesion due to the development of the corrosion process under the film. Therefore, adhesion itself, how high it may be, cannot provide long-lasting protection from corrosion. Likewise coatings with high diffusion limitations, but with poor adhesion cannot provide long-lasting protection [3].

Paint application is one of the most common and reliable ways to protect metal surfaces from corrosion and give a decorative surface finish. It is known that the testing of protective properties of the coating in an operational environment takes a lot of time that does not comply with any developers or manufacturers. Rapid tests provide information about the resistance of the coating under its compulsory destruction simulating natural mechanism of aging in a short trial time. Electrochemical methods are used as these accelerated test methods.

Electrochemical methods are based on measurement of electrical parameters of the electrochemical phenomena occurring in test solution. This measurement was performed using an electrochemical cell, which is a container with the test solution in which electrodes are placed. Electrochemical processes are accompanied by the appearance in the solution or changing of the potential difference between the electrodes or changing of the magnitude of current flowing through a solution [4].

To assess the protective properties and select the modifier concentration in the polymer coating the study of the time dependence of the stationary potential of the metal – coating and removal of the anodic polarization curves was used. Potential Measurement of the metal – coating was being performed in 0.5% HCl at a temperature of $(20 \pm 2)°C$ in the scale of silver chloride reference electrode for 24 h, then values were converted to a scale of the standard hydrogen electrode.

Removal of anodic polarization curves in a 0.5 % HCl was being performed using a potentiostat 50 PI–1, and a driving voltage programmer 8 in potentiostatic mode. Tests were carried out in a three-electrode electrochemical cell. Before removing the anode polarization curve value of the equilibrium potential of the metal – cover was being measured for 5 min. Anodic polarization was carried out in potentiostatic mode at a potential step change in 20 mV with delayed current at each potential for 1 min.

By the slope of the Tafel plot of the polarization curve in the coordinates E–lgi (Fig. 10.1) value of the coefficient b was determined [5]:

$$b = \frac{2,303 \cdot R \cdot T}{\alpha \cdot n \cdot F}.$$

Corrosion current density in the meta-coating was determined graphically by the intersection of stationary potential measured for 24 h (E24) in 0.5% HCI, and the straight portion of the anode polarization curve, which if necessary was extrapolated. It was experimentally established that offset values of stationary potential of corrosive base – covering systems in electronegative side over time may be due to relief of the anode ionization of the metal due to the coating moisture permeability and increase of its conductivity.

For samples with an epoxy polymer coating (EPC) stationary potential takes more electropositive value compared to carbon steel. With increasing concentration of modifying additive in the p- aminophenol coating displacement capacity reaches 30 mV.

For samples with EPC with modifier content of 0.5–5.0% in the potential range −0.2–0.3 Tafel slopes (b) of all areas of the anode curves are approximately the same (Table 10.3). This suggests that the mechanism of active dissolution of iron in the pores of the polymer coating does not change, and inhibition is due to the decrease of the effective surface of the dissolving metal.

Organic modifier p-aminophenol introduced into the coating has two functional groups of atoms containing lone pairs electrons, facilitating its adsorption on iron, which applies to transition metals with a free d- orbitals. The presence of p-aminophenol in the polymer coating decreases the effective surface of the dissolving metal and inhibits the anodic process.

Extrapolation of the linear portions of the polarization curves (Fig. 10.1) to the measured stationary capacity allows to determine the corrosion rate.

Table 10.3 shows the stationary potentials and the corrosion rate of the metal – covering systems.

From these data one can conclude that the polymer coatings inhibit corrosion of carbon steel 08 kp, which corrodes in 0.5% HCl at a speed of 1.95 mA/cm². Application of epoxy polymer coating containing modifier of n-aminophenol 0.5–5.0% reduces the corrosion rate of steel in 1.1–2.2 times.

With increasing concentration of n- aminophenol modifier in the polymer coating corrosion resistance at the system increases whereas the corrosion current density decreases (Fig. 10.2), and the polarization curves are shifted

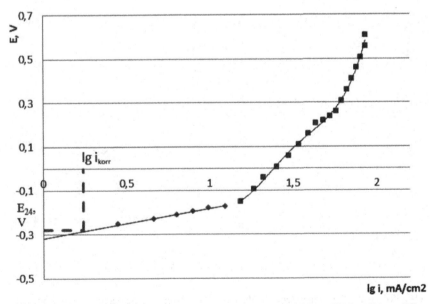

FIGURE 10.1   Anodic polarization curve. Example – Steel 08 kp – EPC + 1% modifier.

TABLE 10.3   Stationary Potentials and the Corrosion Rate of the Steel 08 kp – Epoxy Coating in 0.5% HCl

Tested sample	E, B	lgi	i, mA/cm2	b, B	Percentage of modifier
Carbon steel 08kp	−0.288	0.29	1.95	0.1346	–
Carbon steel 08 kp – LPC + 0.5 M	−0.28	0.24	1.74	0.1130	0.5
Carbon steel 08 kp – LPC + 1.0 M	−0.275	0.21	1.62	0.0978	1
Carbon steel 08 kp – LPC + 2.0 M	−0.266	0.18	1.51	0.0911	2
Carbon steel 08 kp – LPC + 3.0 M	−0.26	0.15	1.41	0.0572	3
Carbon steel 08 kp – LPC + 4.0 M	−0.252	0.11	1.29	0.0434	4
Carbon steel 08 kp – LPC + 5.0 M	−0.238	−0.05	0.89	0.0328	5
tested Sample	E, B	lgi, mA/cm^2	i, mA/cm^2	b, B	Percentage of modifier

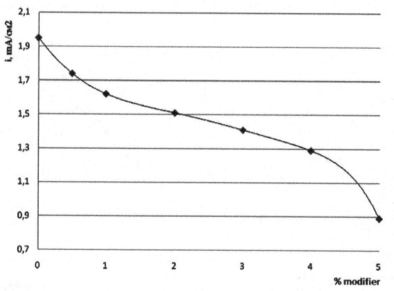

FIGURE 10.2   The dependence of the corrosion current on the percentage of modifier.

to lower currents. Basing on the obtained results it can be concluded that the polymer coatings based on epoxy resins with modifier inhibit corrosion of carbon steel better than without it.

10.3 CONCLUSION

Introduction of n-aminophenol modifier at a concentration of 0.5–5.0 wt. % into a polymeric epoxy resin coating improves the corrosion resistance of the system to 0.5% HCl, while the current density of steel corrosion 08kp decreases from 1.95 to 0.89 mA/cm^2 and the polarization curves are shifted to lower currents. Application of the epoxy polymer coating modifier in an amount of 5% reduces the corrosion rate of carbon steel in 08 kp 0.5% HCl 2.2 times. Addition of 0.5–2.0% modifier also improves physical and mechanical characteristics (improved strength, hardness and adhesion of coatings).

Experimental studies have shown that additional evaluation of protective properties of coatings by electrochemical methods in conjunction with conventional for the paint industry methods of research yielded a better understanding of corrosion processes under cover, allowed to assess their impact on the course of the concentrations of the modifier. These studies allowed to adjust the coating composition to the individual applications in order to achieve a high degree of protection of metal surfaces.

KEYWORDS

- aminophenol
- coatings
- corrosion
- electrochemical method
- epoxy polymer
- mechanical properties
- oligoepoxyde

REFERENCES

1. Li, H. Handbook of epoxy resins. H. Lee and K. Neville. Moscow: Energiya, 1973. 268 p.
2. Kireyeva, V. G. Coating materials based on epoxy resins. V.G. Kireyeva. M.: NII-TEKHIM, 1992. 48 p (Survey information. Ser "Paint industry").
3. Rosenfeld, I. L. Corrosion Inhibitors. I.L. Rosenfeld. – Moscow: Khimiya, 1977. 352 p.
4. Semenova I. V. Corrosion and Corrosion Protection. I.V. Semenova, G.M. Florianovich, A.V. Khoroshilov; ed. I.V. Semenova. Moscow: Fizmatlit 2006. 328 p.
5. Damascus B. B., Peter O. Electrochemistry: A Textbook for Chem. Faculty Universities. Moscow: Higher School, 1987, 295 p.

CHAPTER 11

MODIFICATION OF POLYAMIDE-6 BY N,N'-BIS-MALEAMIDOACID

N. R. PROKOPCHUK,[1] E. T. KRUT'KO,[2] and
M. V. ZHURAVLEVA[3]

[1]Corresponding Member of Belarus NAS, Professor, Head of
Department (BSTU); [2]D.Sc. (Engineering), Professor (BSTU)
[3]Professor (BSTU); PhD student (BSTU); Belarusian State
Technological University, Sverdlova Str.13ª, Minsk, Republic of
Belarus, E-mail: v.polonik@belstu.by

CONTENTS

ABSTRACT

This chapter is devoted to receiving and research of compositions on the basis of industrially made polyamide-6 modified by N,N'-bis-maleamidoacid of meta-phenylenediamine. Existence in macromolecules of polyamides reactive carboxyl, amide and amino groups, capable to interact with multifunctional monomeric and oligomeric modifiers, gives the chance of receiving the materials possessing properties of sewed polymers. As a result the use of bis-amido acid as a modifying additive in the system of aliphatic polyamide-6 provides the improvement of strength properties and thermal characteristics of a polymeric material.

11.1 INTRODUCTION

Due to the ever increasing demands of engineering industries of new techniques to polymeric materials, including created on the base of aliphatic polyamides, an urgent task of improving performance properties of these polymers, including high heat and heat resistance, resistance to thermal-oxidative and aggressive media and adhesion characteristics to various substrates remains actual. Physical properties of aliphatic polyamides are mainly conditioned by strong intermolecular interactions at the expense of hydrogen bonds that are formed between the amide groups of neighboring macromolecules. In the preparation of polyamides it is necessary to enter regulators (stabilizers) of molecular weight, in this case acetic, adipic acids (as the cheapest and most available) and amines, salts of monocarboxylic acids and monoamines or N-alkyl amides of monocarboxylic acids are used [1]. Polyamides as well as other polymers are polydispersed. In the polymer composition there are large macromolecules and low molecular weight amides with a small number of elementary units in the molecules (oligomers). They contain terminal amino groups and carboxyl groups. Furthermore, the oligomers of amides may also exist in a cyclic form. The mechanism of their formation is still not clear. However, it is experimentally proved that in the absence of water cyclic oligomers are not capable of polymerization, but they are easily converted into a polyamide in the presence of water. In this regard, the use of additives that at processing of polyamides may release small amounts of moisture can be very useful for improving the performance properties of materials, products and coatings based on polyamides. One of the effective ways of targeted regulation of properties of commercially available polyamides is their chemical modification by polyfunctional reactive compounds [2–5]. Presence in the macromolecules of polyamides reactive carboxyl, amide and aminogroups which are capable to react with monomeric and oligomeric polyfunctional modifiers enables the production of materials having the properties of cross-linked polymers. Thus, it was found that imido compounds, in particular N,N'-bis-maleimides are effective modifiers of many polymers, also including polyamides [6]. Information about the use of N,N'-bis maleinamido acids as the modifiers in their synthesis of intermediates in the scientific literature were not found.

11.2 EXPERIMENTAL

The aim of this work is to study and research the compositions based on polyamide 6 (PA-6), industrially produced by JSC "Grodnoazot" (Grodno, Belarus) (OST 6–06–09–93), intermediate product in the synthesis of modified-N,N'-metaphenylene bis maleineimide (FBMI)-N,N'-bis-maleamido metaphenylenediamine (FBMAK)

The modifying reagent was injected into the PA-6 at doses of 5–10 wt. %. FBMAK choice, as previously FBMI [6], due to their high reactivity associated with the content of the reactant molecules of double bonds that can be disclosed under thermal or photographic processing to form a spatial cross-linked polymer structure [6]. Physical and mechanical properties of PA-6 are shown in Table 11.1.

Receiving of FBMAK was performed by reaction of equimolar amounts of metaphenylenediamine with maleic anhydride at 20–25°C by gradual addition to a solution of diamine in a minimum amount of solvent – dimethylformamide stoichiometric amounts of maleic anhydride. To obtain the corresponding bis-maleinimide, the second step (imidization) of bis-amidoacid by heating it in a mixture of acetic anhydride and

TABLE 11.1   Physical and Mechanical Properties of PA-6

Index name	Norm on the highest quality category
Color	From white color to light yellow color
Number of point inclusions per 100 g of product units	Not more than 8
Size of crumbs msm,	1.5–4.0
Moisture content, %	0.03–2.00
Relative viscosity, dl/g	2.20–3.50
Mass fraction of extractable, %	1.0–3.0
Melting point, °C	214–220

sodium acetate in the ratio 2.5:0.5 moles per mole of bis-amidoacid at 70–90°C was performed. Upon completion of imidization (about 2.5 h of heating) FBMI was isolated and recrystallized from a mixture of ethyl and n- propyl alcohol, in the ratio 1:1. It should be noted that the yield of the intermediate product − FBMAK − in the first synthesis step is 85−90% while the final product after imidization (FBMI) is only 50−60%.

The melting point of the synthesized N,N'-metaphenylene-bis-maleinimide was 203°C, which corresponded to the literature data [7]. FBMAK when heated in the process of determining the melting point and become FBMI and the melting point of this compound is not clearly fixed.

Receiving of FBMAK and FBMI is performed according to Scheme 1. It is important to note that the perspective of use of polyamide 6 FBMAK as a modifier component instead of FBMI allows to exclude out of the process of synthesizing FBMI the step of high-cost chemical cyclodehydration (imidization) of FBMAK, moreover, the loss of the product during its conversion to the bis-imide of the final step of synthesis is eliminated, and during imidizing process extracted water promotes destruction of cyclic structures in the system of the polyamide.

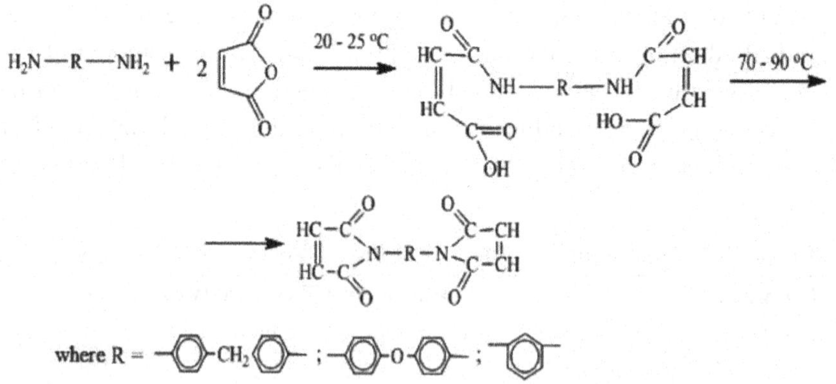

SCHEME 1

To study the structure of polyamide-6, the processes occurring in the polymer system, as well as assessing the completeness of spending FBMAK reactive groups by reacting with amine and amide functional groups of PA-6 during formation of three-dimensional structure when heated samples was performed IR spectroscopic study using FTIR spectrometer Nicolet 7101 (USA) in the range of 4000–300 cm^{-1} (resolution 1 cm^{-1}) [8]. Furthermore, the possibility of formation of intermolecular cross-linking was confirmed by electron paramagnetic resonance (EPR) [9, 10]. EPR spectra are recorded using the modified RE −1306 spectrometer with computer software. The heating process of polyamide compositions of modified FBMAK was carried out in the cavity of the spectrometer in the temperature range 20−250°C. Before recording the spectra the samples were cooled to room temperature. MnO containing ions of Mn was used as an external standard [9] nitroxyl radicals were used as markers.

For example, evidence of more efficient structuring system in polyamide compositions containing as builders FBMAK compared with FBMI is that the time of correlation of stable nitroxyl radicals introduced into FBMAK modified PA-6, increased from $23·10^{-10}$s (source unmodified P6) to $50·10^{-10}$ s with polyamide-6, modified by 5 wt % FBMAK. For PA-6, modified with the same amount of MFBI correlation time of nitroxide was significantly shorter period from $40·10^{-10}$ s.

Comparative analysis of the IR absorption spectra obtained in an airless environment of PA-6 compositions and FBMAK different composition (5−10 wt. % FBMAK builder) in the polymer melt, followed by heating of the samples at 150–210°C showed that for the modified polymer systems absorbance decrease of absorption bands in the 1647 cm^{-1}, characteristic of the amide groups is observed.

It appears that at the process of chemical modification of PA-6 by bis-amidoacid, as well as modifying it by its corresponding bis-imide reactions, resulting in an increase in the molecular weight of PA-6 take place. It occurs as by the interaction of the carboxyl groups of the bis-amidoacid modifier with amine-terminated polyamide groups (Scheme 2) and by the formation of interchain crosslinks (Scheme 3) by opening the double bonds formed during the heat treatment at 150−200°C samples of polyamide compositions. This does not prevent homopolymerization of FBMI and that is due to the ability of the double bonds in FBMI to be activated by adjacent carbonyl groups of the imide cycle and disclosed by reacting with compounds containing labile hydrogen atom.

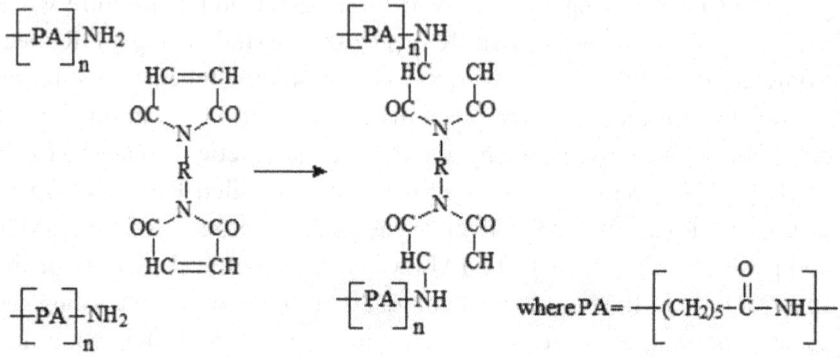

$$\left[PA\right]_n NH_2$$

HC=CH
OC   CO
  N
  R
  N
OC   CO
HC=CH

$$\left[PA\right]_n NH_2$$

$$\longrightarrow$$

$$\left[PA\right]_n NH$$

HC   CH
OC   CO
  N
  R
  N
OC   CO
HC   CH

$$\left[PA\right]_n NH$$

where PA = $\left[(CH_2)_5-\overset{O}{\overset{\|}{C}}-NH\right]$

SCHEME 2

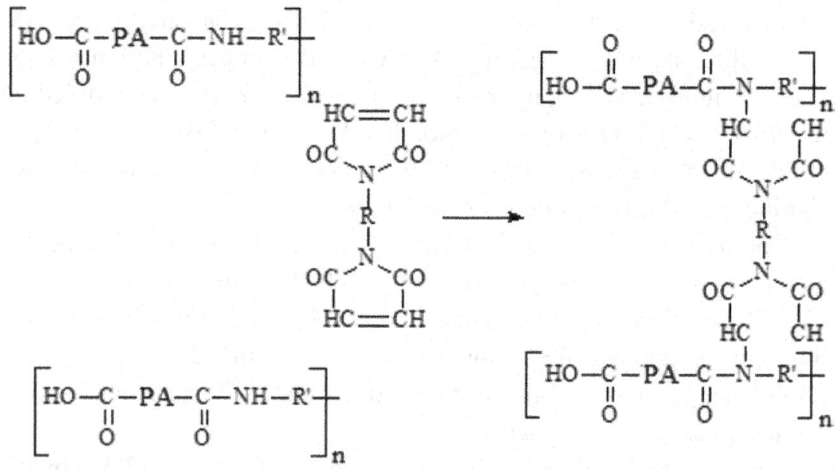

$$\left[HO-\overset{O}{\underset{\|}{C}}-PA-\overset{O}{\underset{\|}{C}}-NH-R'\right]_n$$

HC=CH
OC   CO
  N
  R
  N
OC   CO
HC=CH

$$\left[HO-\overset{O}{\underset{\|}{C}}-PA-\overset{O}{\underset{\|}{C}}-NH-R'\right]_n$$

$$\longrightarrow$$

$$\left[HO-\overset{O}{\overset{\|}{C}}-PA-\overset{O}{\overset{\|}{C}}-N-R'\right]_n$$

HC   CH
OC   CO
  N
  R
  N
OC   CO
HC   CH

$$\left[HO-\overset{O}{\underset{\|}{C}}-PA-\overset{O}{\underset{\|}{C}}-N-R'\right]_n$$

SCHEME 3

These processes altogether result in a change of the supramolecular structure of the PA-6, causing the improvement of mechanical, thermal and adhesive properties of the polymer [6].

In practical use of FBMI their thermochemical characteristics can have essential significance. In this connection it is necessary and appropriate to examine the thermochemical characteristics of a number of different structures of BMI by differential scanning calorimetry.

DSC curves (Fig. 11.1) of all studied BMI taken in the temperature range of 20–400°C at a rate of temperature rise of 10°C/min are clearly

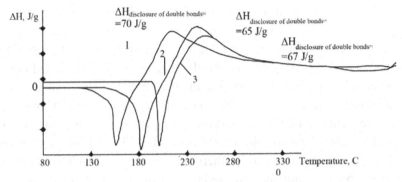

FIGURE 11.1 DSC curves of BMI. 1 – N,N'-diphenylmethane-bis-maleinimide; 2 – N,N'-diphenyloxide-bis-maleinimide; 3 – N,N'-m-phenylene-bis-maleinimide (FBMI)

played two peaks endothermic the melting process of BMI and exothermic corresponding to the process of disclosure of the double bonds.

The position of the peaks in the respective process on the temperature scale depends on the chemical structure of compounds. According to the value of enthalpy disclosure of double bonds, studied compounds in their thermochemical characteristics and probably related to reactivity in the reactions occurring with the opening of the double bonds are not significantly different. In this connection, the use of aliphatic polyamides FBMAK, FBMI is justified as modifying component. This component (MFBI), and accordingly MFBAK as its intermediate are produced in industrial conditions and are widely used for the production of a large range of heat-resistant composite materials. They are used to improve the adhesion of the cord to rubber mass in the manufacture of various kinds of tires for the automotive industry.

Effectiveness of modifying action of FBMAK in thermo-oxidative degradation processes of PA-6 compositions was evaluated on samples of three types: model modified FBMAK powder compositions, films, molded samples.

Model modified powder compositions of PA-6 were prepared for the rapid analysis of possible stabilizing action FBMAK to the polymeric matrix. They were prepared as follows: PA-6 unstabilized granules were placed in a mixer, where nitrogen was supplied to provide oxygen free environment. The temperature was raised to 240°C and a stirrer was switched on. With continuous stirring, the calculated amount of FBMAK was injected into the polymer melt. The mixture was being stirred for 5 min. Then it was poured into a mold and cooled. After curing, the sample was removed from the mold and crushed at a cryogenic temperature (after soaking for 10 min in liquid nitrogen at −180°C) to a fine powder. Investigation of the thermal stability was carried out by differential thermal and thermogravimetric analysis. The sample weight was 10 mg, a heating rate in the range of 20−500°C was 5°C/min. The activation energy of thermal oxidative degradation of the polymer Ed was calculated from the Broido method [11−13] according to dynamic thermogravimetry obtained by thermo-setting − module TA-400 of the Mettler Toledo company (Switzerland) in the temperature range 330−400°C. Instrument calibration was performed on indium standard.

Laboratory samples of the polyamide films of 15−20 microns thick were cast from solutions of PA-6 with formic acid. To obtain solutions a sample of the polymer (2 g) with a relative viscosity of 2.8 (determining in a solution of concentrated sulfuric acid) was dissolved in 28 mL of formic acid at room temperature. After that the estimated quantity of FBMAK was injected into the solution and mixed thoroughly. Then degreased and dried on glass plates by pouring through a nozzle with an adjustable gap the layer of film-forming composition was deposited. The glass plates were placed horizontally in a vacuum drying oven, where at a residual pressure of 2 mm Hg. Art. and a temperature not above 30°C for 4.5 h the solvent was removed. Formed on a glass substrate polyamide film was removed and subjected to research.

Samples were also prepared by injection molding as follows: FBMAK modifier in an amount of 5−10 wt. % by weight of PA-6 was applied to the polymer granules without the heat stabilizer with a relative viscosity of 2.8 by dusting. Powdered pellets were peppered through the hopper to the injection-molding machine, where they were melted. From the molten mass bilateral blades with working part 50 × 10 × 3 mm were cast. Experimental samples one day after their manufacture were placed in a heat chamber with a free circulation of air. They were maintained at 150°C for predetermined time (30−210 h). Determination of the breaking strength of the samples was performed by stretching according to GOST 112−80 using the tensile testing machine T 2020 DS 10 SH (Alpha Technologies UK, USA) at room temperature and velocity of the gripper 100 mm/min. Judging by the results of measurements of the tensile strength of the samples given in Table 11.2 generated in the step of heat treating of polyamide

TABLE 11.2 Tensile Strength of PA-6 and Modified by MFBI MFBAK Samples of PA-6 After Thermal Oxidation in Air at 150°C for 30–210 h

Material	Tensile strength, MPa								T_d, °C	E_d, kJ/mol
	initial	after thermal oxidation in air for, h								
		30	60	90	120	150	180	210	30	30
PA-6	59.0	50.0	49.0	41.0	35.5	30.5	23.0	18.6	275	120
PA-6 + 5 wt. % FBMAK	64.0	66.5	64.0	61.4	61.0	61.5	61.0	60.0	285	135
PA-6 + 10 wt. % FBMAK	68.0	69.0	69.5	62.4	62.0	63.4	63.0	61.0	290	148
PA-6 + 5 wt. %FBMI [6]	63.4	64.6	63.0	62.6	62.0	62.0	61.4	60.7	287	141

compositions meshwork of aliphatic polyamide-6 at the expense of the reactive groups of FBMAK (content in the composition of 5–10 wt.%) and functional groups of PA-6 predetermines the higher rates of heat resistance (temperature of thermo-oxidative degradation increases by 10–15°C) and deformation-strength properties (tensile strength increases by almost 10 MPa at initial modified samples, amounting 64.0–68.0 MPa) as compared to the unmodified polyamide 6 (59.0 MPa) and is comparable with PA-6 modified by the bis-maleinimide (FBMI).

Importantly, FBMAK as well as FBMI, provides a thermally stabilizing effect on polyamide-6 at various temperature and time of exposure of samples of polymer compositions. Calculation of the activation energy of thermal oxidative degradation of the samples (Ed) also indicates the formation of crosslinks in a matrix structure of polyamide 6 at the expense of additives of MFBAM and MFBI, increasing depending on the modifier content of FBMAK from 135 to 148 kJ/mol.

As seen from Table 11.2 polyamide 6 modified with bis maleamidoacid has not only improved strength characteristics as compared with the original PA-6, but it is also more resistant to high temperature and time fields.

11.3 CONCLUSION

Thus, the use of bis-aminoacid (FBMAK) as a builder in the system of aliphatic polyamide-6 improves strength properties and thermal characteristics of a polymeric material based on PA-6 virtually in the same range as the previously used for the same purpose FBMI [6].

KEYWORDS

- energy activation
- modification
- N,N'-diphenylmethane-bis-maleinimide
- N,N'-diphenyloxide-bis-maleinimide
- N,N'-m-phenylene-bis-maleinimide
- polyamid-6

REFERENCES

1. Kudryavtsev, G.I., Noskov, M.P., Volokhina, A.V. Polyamide fibers. M.: "Chemistry." 1976. 259 p.
2. Hapugalle, G., Prokopchuk, N.R., Prakapovich, V.P., Klimovtsova, I.A. New thermostabilizers polyamide-6. NASB Vestsi. Ser.him. navuk. 1999. № 1, 114–119.
3. The anionic polymerization of ε-caprolactam and its copolymerization with ω-dodekalaktamom in the presence of aromatic polyimides. Y.S. Vygodskii et al. Polym. comp. Ser. A. 2006. 48(6), 885–891.
4. The anionic polymerization of ε-karolaktama in the presence of aromatic polyimides as macromolecular activators. Y.S. Vygodskii et al. Polym. Comp. Ser. A. 2003. T. 45, № 2. P. 188–195.
5. The anionic polymerization of ε-caprolactam in the presence of aromatic diimides. Y.S. Vygodskii et al. Polym. comp. Ser. A. 2005. Volume 47, № 7. P. 1077.
6. About modifying the action of N,N'-bis-imides of unsaturated dicarboxylic acids, aliphatic polyamides. V. Biran et al. Reports of the Academy of Sciences of the Byelorussian SSR. 1983. V. 27, № 8. P. 717–719
7. Synthesis of N,N'-bis-imides of unsaturated dicarboxylic acids, cycloaliphatic. A.I. Volozhin et al. Vestsi BSSR. Ser. him. Navuka. 1974. № 1. P. 98–100.
8. Nakanishi, K. Infrared spectra and structure of organic compounds. K. Nishi Naka. New York: Wiley, 1965. P. 216.
9. The study by EPR of free radicals in polyimides. A.G. Boldyrev et al. Report of the Academy of Sciences of the USSR. 1965. T. 163, № 5. P. 1143–1146.
10. Ingram, D. Electron Paramagnetic Resonance in Biology. D. Ingram; under Society. Ed. Y.I. Agip, L.P. Kayushin. New York: Wiley, 1972. P. 296.
11. Broido A., Semple A. Sensitive graphical method of treating thermogravimetric analysis date. I. Polym. Sci. 1969. Part A. 2. v. 7. № 10. P. 1761–1773.
12. Prokopchuk N.R. study the thermal stability of polymers by derivatography. Vestsi BSSR. Ser. him. Navuka. 1984. № 4. 119 –121.
13. Polymer construction. Method for determining the durability of the activation energy for thermal oxidative degradation of polymeric materials: STB 13330–2002. Enter. 28.06.2002. Minsk: Technical Committee on technical regulation and standardization in construction: the Ministry of Architecture and Construction of Belarus, 2002. 8 p.

CHAPTER 12

EFFICIENT SYNTHESIS OF A BISSPIRO-3,1-BENZOXAZIN-4,1'-CYCLOPENTANES

SH. M. SALIKHOV,[1] R. R. ZARIPOV,[2] S. A. KRASKO,[3]
L. R. LATYPOVA,[1] N. M. GUBAIDULLIN,[2] and
I. B. ABDRAKHMANOV[1]

[1]Institute of Organic Chemistry Ufa Scientific Centre of Russian Academy of Sciences, Prospect Oktyabrya 71, 450054, Ufa, Russia, Tel: +7 (347) 235 55 60. E-mail: Salikhov@anrb.ru

[2]Bashkir State Agrarian University, 50 Let Oktyabrya, 21, 450001, Ufa, Russia, Tel: +7 (347) 235 55 60

[3]Ufa State Petroleum Technological University, Kosmonavtov 1, 450062, Ufa, Russia, Tel: + 7 (347) 242 09 35. E-mail: ksa.85@mail.ru

CONTENTS

ABSTRACT

The new data on the directed synthesis of bisspiro-3,1-benzoxazine based on dicarboxylic acid 2-(1-cyclopentenyl) aniline. The influence of various substituents in the benzene ring of ortho siklopentana on the reaction of their intramolecular heterocyclization.

12.1   INTRODUCTION

Benzoxazines are the most promising compounds for the synthesis of a new generation of organic substances with pronounced biological activity. Major effective and widely used approaches to create 3,1-benzoxazines based on the classical condensation methods of amines with carboxylic acids derivatives and further cyclization of products by various regents and catalysts [1]. Previously, we have developed a method of the available synthesis of 3,1-benzoxazine, which allowed to include amino-Claisen rearrangement products in this process [2]. Initiating systems based on these compounds and their metal complexes have found practical substantiation as polymer molecular weight regulators in controlled radical polymerization [3] and antioxidants of radical chain oxidation of polyisobutylene.

In order to expand the combinatory library of 3,1-benzoxazines by the proposed approach [4] we continue studying synthetic capacity of available dicarboxylic acids.

So, the reaction of chloranhydride with the ortho-(cyclopent-1-enyl)-aniline **1** in the presence of K_2CO_3 led the amides **2–8**. The reaction of compounds **5–8** with $HCl_{(g)}$ and the subsequent treatment with 10% $NaHCO_3$ gives benzoxazines **9–12**. In the case of amides **2–4** formation of 3,1-benzoxazine cycle is not observed. Probably insignificant length of methylene bridge is a steric hindrance to the formation of two 3,1-benzoxazinone cycles at a short distance from each other.

13 R = 2-N O₂	19 R = 2-N O₂
14 R = 3-N O₂	20 R = 3-N O₂
15 R = 2-O CH₃	21 R = 2-O CH₃
16 R = 2-I	22 R = 2-I
17 R = 3-I	23 R = 3-I
18 R = 2-C l 4-Cl	24 R = 2-C l 4-Cl

Further the range of 3,1-benzoxazine synthesized from monocarboxylic acids was extended. It is established that the chloroanhydrides of benzoic acids react with ortho-(cyclopent-1-enyl)aniline **1** at mild conditions to form amides **13–18** in high yields. The treatment of these amides by $HCl_{(g)}$ led to benzoxazines **19–24**.

Benzoxazines based on pyridinecarboxylic acids, where the introduction of an electron-deficient pyridine ring favors the formation of the conjugated system in which there are two coordinating center to form a donor-acceptor bond with ions of metals are also of practical interest. Benzoxazin **26** was obtained from amide **25** using a method described above.

Intramolecular heterocyclization of arylamides **A** to 3.1-benzoxazines is initiated by the proton-catalyzed addition of cyclopentadiene ring to the double bond and the generation of carbenium ion of benzyl-type **B**; the subsequent intramolecular stabilization ions by nucleophilic oxygen atom of an amide fragment gives heterocyclic ions **C** – precursors of neutral products of the rearrangement **D**. Since, the cyclization reaction is probably limiting stage, the introduction of donor substituents into the benzene ring of the initial substrate promotes the intramolecular heterocyclization of ortho-cyclopentanedione. The cyclization of amides under the electron impact mass spectrometry conditions [5] confirmed this fact.

12.2 EXPERIMENTAL SECTION

Spectra of ^1H and ^{13}C NMR were recorded on Bruker AM-300 (300.13 and 75.47 MHz) and Bruker Avance III 500 (500.13 and 125.75 MHz), using $CDCl_3$, and acetone-d6 as solvents. Chemical shifts are reported in units of parts per million and all coupling constants are reported in hertz. All reaction were monitored by TLC analysis on plates "Sorbfil PTLC-A-AF."

12.2.1 GENERAL SYNTHETIC PROCEDURE CARBOXYLIC ACID AMIDES

To a solution of chloroanhydride (0.014 mol) in CH_2Cl_2 under stirring at room temperature ortho-(cyclopent-1-enyl)-aniline (0.01 mol) and K_2CO_3 (0.02 mol) was added. The reaction mixture was stirred for 24 h. After completion of the reaction (control by TLC), the precipitate was filtered off, the filtrate was washed with water (1 × 25 mL), 10% $NaHCO_3$ (2 × 25 mL), dried over $MgSO_4$. The solvent was evaporated, the residue was purified by column chromatography (petroleum ether: ethyl acetate).

N,N'-bis(2-cyclopent-1-en-1-ylphenyl)ethanediamide (2): white solid (54%); mp 202–204°C. ^{1}H NMR (CDCl$_3$ +acetone – d$_6$): 1.90 m (2H, C^{4}H), 2.46 m (2H, C^{5}H), 2.54 m (2H, C^{3}H), 5.84 t (2H, J = 2.0, C^{2}H), 6.97 – 7.08 (3H, m, Ar), 8.19 (1H, d, J = 7.3, Ar), 9.73 (1H, br, NH). ^{13}C NMR (CDCl$_3$ +acetone – d$_6$): 22.96 (C^4), 33.53 (C^5), 36.33 (C^3), 119.66 (C^2), 124.54 (Ar), 127.23 (Ar), 127.69 (Ar), 128.79 (C^1), 131.27 (Ar), 132.72 (Ar), 139.22 (Ar), 156.95 (C=O).

N,N'-bis(2-cyclopent-1-en-1-ylphenyl)malonamide (3): orange oil (40%); ^{1}H NMR (CDCl$_3$): 2.0 (2H, m, C^{4}H), 2.18 (1H, s, COCH$_2$CO), 2.53 (2H, m, C^{5}H), 2.71 (2H, m, C^{3}H), 6.0 (1H, J = 2.0, t, C^{2}H), 6.75–7.15 (4H, m, Ar). ^{13}C NMR (CDCl$_3$): 23.23 (C^4), 33.47 (CH$_2$), 33.98(C^{5}H), 36.44 (C^{3}H), 117.38 (Ar), 120.23 (Ar), 125.62 (C^{2}H), 127.66 (Ar), 128.35 (Ar), 129.41 (C^{2}H), 135.84 (Ar), 140.51 (Ar), 140.93 (C=O).

N,N'-bis(2-cyclopent-1-en-1-ylphenyl)succinamide (4): yellow oil (50%). ^{1}H NMR (CDCl$_3$): 1.85 (2H, m, C^{4}H), 2.34 (2H, m, C^{5}H), 2.48 (2H, m, C^{3}H), 2.77 (2H, s, CH$_2$), 5.57 (1H, J = 2.0, t, C^{2}H), 7.0–7.30 (4H, m, Ar). ^{13}C NMR (CDCl$_3$): 23.74 (C^{4}H), 28.59(CH$_2^1$), 33.59 (C^{5}H), 35.78(C^{3}H), 126.48(C^2), 127.84 (Ar), 128.62 (Ar), 129.19 (Ar), 129.30 (C^1), 129.36 (Ar), 137.15 (Ar), 140.45 (Ar), 176.54 (C=O).

N,N'-bis(2-cyclopent-1-en-1-ylphenyl)pentanediamide (5): yellow oil (66%). ^{1}H NMR (CDCl$_3$): 1.81 (1H, m, CH$_2$), 2.02 (2H, m, C^{4}H), 2.49 (2H, m, C^{5}H), 2.57 (2H, m, C^{2}H), 2.70 (2H, m, CH$_2$), 5.88 (1H, J=1.8, t, C^{2}H), 6.73 (1H, td, J=1.6, 7.6, C^{4}H), 7.10–7.25 (4H, m, Ar), 8.25 br. (1H, NH). ^{13}C NMR (CDCl$_3$): 22.99 (CH$_2$), 23.20 (C^{4}H), 33.69 (C^{3}H), 36.14 (C^{5}H), 36.58 (CH$_2$), 115.70 (Ar), 118.10 (Ar), 127.57 (Ar), 128.13 (Ar), 128.55 (C^1), 130.47 (C^2), 134.37 (Ar), 140.47 (Ar), 170.43 (C=O).

N,N'-bis(2-cyclopent-1-en-1-ylphenyl)hexanediamide (6): yellow oil (38%). ^{1}H NMR (CDCl$_3$): 1.19 (4H, s, CH$_2$), 2.02 (2H, m, C^{4}H), 2.48 (2H, m, C^{5}H), 2.61 (2H, m, C^{2}H), 2.76 (4H, s, CH$_2$), 5.87 (1H, J = 2.0, t, C^{2}H), 6.60–7.07 (4H, m, Ar). ^{13}C NMR (CDCl$_3$): 23.23 (C^{4}H), 29.73(C^{3}H), 30.95 (C^{5}H), 33.83 (CH$_2$), 36.63(CH$_2$), 123.75 (C^1), 125.31 (Ar), 126.82 (Ar), 127.53 (Ar), 125.74(C^2), 127.97 (Ar), 128.29 (Ar), 141.32 (Ar), 146.57 (Ar), 161.13 (C=O).

N,N'-bis(2-cyclopent-1-en-1-ylphenyl)isophthalamide (7): white solid (68%). mp. 105–107°C. ^{1}H NMR (CDCl$_3$): 1.95 (4H, m, C^{11}H$_2$, C^{11a}H$_2$), 2.50 (4H, m, C^{10}H$_2$, C^{10a}H$_2$), 2.72 (4H, m, C^{9}H$_2$ C^{9a}H$_2$), 6.03 (2H, J = 2.1, t, C^{8}H, C^{8a}H), 7.15 (2H, t, J = 7.4, C^{4}H, C^{4a}H), 7.27 (2H, d, J = 7.9, C^{6}H, C^{6a}H), 7.32 (2H, t, J = 7.4, C^{5}H, C^{5a}H), 7.68 (1H, t, J = 7.7, C$^{5'}$H), 7.95 (2H, d, J = 6.9, C^{6}H, C^{6a}H), 8.15 (2H, d, J = 7.9, C^{6}H, C$^{4'}$H), 8.50 (1H, s, C$^{2'}$H), 9.45 (2H, br, NH). ^{13}C NMR (CDCl$_3$): 23.27 (C^{10}, C^{10a}), 33.78 (C^{11}, C^{11a}), 36.77 (C^9, C^{9a}), 121.25 (C^6, C^{6a}), 124.28 (C^4, C^{4a}), 125.64 (C$^{2'}$), 127.65 (C^8, C^{8a}), 127.72 (C^6, C$^{4'}$), 129.02 (C^1, C^{1a}), 129.28 (C$^{5'}$), 129.74 (C^5, C^{5a}), 130.86 (C^3, C^{3a}), 134.31 (C^2, C^{2a}), 135. 52 (C^7, c^{7a}), 140.77 (C$^{1'}$ C$^{3'}$), 163.72 (C=O).

N,N'-bis(2-cyclopent-1-en-1-ylphenyl)terephthalamide (8): white solid (44%). mp. 205–207°C. ^{1}H NMR (CDCl$_3$): 2.29 (4H, m, C^{11}H$_2$, C^{11a}H$_2$), 2.45 (4H, m, C^{10}H$_2$, C^{10a}H$_2$), 2.84 (4H, m, C^{9}H$_2$, C^{9a}H$_2$), 5.69 (2H, t, J = 2.1, C^{9}H, C^{8a}H), 6.42 (2H, t, J = 7.4, C^{4}H, C^{4a}H), 7.27 (2H, d, J = 7.9, C^{6}H, C^{6a}H), 7.40 (2H, t, J = 7.4, C^{5}H, C^{5a}H), 7.44 (1H, t, J = 7.7, C$^{5'}$H), 7.94 (2H, d, J = 6.9, C^{6}H$_2$, C^{6a}H), 8.31 (2H, d, J = 7.9, C^{6}H, C$^{4'}$H), 9.19 (2H, br, NH). ^{13}C NMR (CDCl$_3$): 25.89(C^{10}, C^{10a}), 27.54(C^{11}, C^{11a}), 29.66(C^9, C^{9a}), 119.39(C^6, C^{6a}), 124.39(C^4, C^{4a}), 126.51(C$^{2'}$), 127.31(C^8, C^{8a}), 127.83(C^6, C$^{4'}$), 131.85(C^2, C^{2a}), 135.13(C^7, C^{7a}), 141.26(C$^{1'}$ C$^{3'}$), 161.07(C=O).

N-(6-cyclopent-1-en-1-ylcyclohexa-1,5-dien-1-yl)-2-nitrobenzamide (13): yellow solid (65%). mp. 91–93°C. ^{1}H NMR (CDCl$_3$): 1.95 (2H, m, H$^{5'a}$, H$^{5'b}$), 2.48 (2H, m, H$^{4'a}$, H$^{4'b}$), 2.67 (2H, m, H$^{3'a}$, H$^{3'b}$), 5.91 (1H, t, ^{3}J = 1.8, H$^{2'}$), 7.15 (1H, t, ^{3}J$_{4-3}$ = 7.3, ^{3}J$_{4-5}$ = 7.3, H^4), 7.22 (1H, d, ^{3}J$_{3-4}$ = 7.3, H^3), 7.31 (1H, t, ^{3}J$_{5-6}$ = 7.3, ^{3}J$_{5-4}$ = 7.3, H^5), 7.62 (2H, m, H°, H^p), 7.72 (1H, t, ^{3}J$_{m-p}$ = 7.4, ^{3}J$_{m-o}$ = 7.4, H^m), 8.11 (1H, d, ^{3}J$_{m'-p}$ = 8.1, H$^{m'}$), 8.33 (1H, d, ^{3}J$_{6-5}$ = 7.3, H^6). ^{13}C NMR (CDCl$_3$): 23.31 (C$^{5'}$), 33.89 (C$^{4'}$), 36.78 (C$^{3'}$), 121.73 (C^6), 124.85 (C^4), 124.89 (C$^{m'}$), 127.86 (C^5), 127.87 (C^3), 128.26 (C°), 130.82 (C$^{2'}$), 130.89 (C^p), 132.51 (C^1), 133.08 (C$^{p'}$), 133.99 (C^m), 140.59 (C^2), 146.50 (C$^{o'}$), 164.14 (C=O).

N-(6-cyclopent-1-en-1-ylcyclohexa-1,5-dien-1-yl)-3-nitrobenzamide (14): yellow solid (68%). mp. 94–95°C. ^{1}H NMR (CDCl$_3$): 2.09 (9H, m, H$^{5'a}$, H$^{5'b}$), 2.66 (2H, m, H$^{4'a}$, H$^{4'b}$), 2.73 (2H, m, H$^{3'a}$, H$^{3'b}$), 6.0 (1H, t, ^{3}J = 1.8, H$^{2'}$), 7.15 (1H, dd, ^{3}J$_{4-3}$ = 7.3, ^{3}J$_{4-5}$ = 7.3, H^4), 7.25 (1H, d, ^{3}J$_{3-4}$ = 7.3, H^3), 7.31 (1H, d, ^{3}J$_{5-4}$ = 7.3, ^{3}J$_{5-6}$ = 7.3, H^5), 7.71(1H, dd, ^{3}J$_{m'-p}$ = 7.9, ^{3}J$_{m'-o'}$ = 7.9, H-m'), 8.23 (1H, d, ^{3}J$_{o'-m'}$ = 7.9, H$^{o'}$), 8.40 (1H, d, ^{3}J$_{p-m'}$ = 7.9, H^p), 8.43 (1H, d, ^{3}J$_{6-5}$ = 7.3, H^6), 8.61 (1H, br, NH), 8.63 (1H, s, H$^{o'}$). ^{13}C NMR (CDCl$_3$): 23.41 (C^5), 24.75 (C$^{5'}$), 33.91 (C^4), 37.09 (C^3), 120.08 (C^6), 121.51 (C^o), 134.62 (C^4), 126.30 (C^o), 127.86 (C^3), 127.99 (C^5), 128.85 (C$^{1'}$), 130.26 (C$^{m'}$), 131.31 (C$^{o'}$), 134.09 (C^1), 136.61 (C$^{p'}$), 141.11 (C^2), 148.32 (C^m), 162.26 (C=O).

N-(6-cyclopent-1-en-1-ylcyclohexa-1,5-dien-1-yl)-2-methoxybenzamide (15): brown oil (79%). ^{1}H NMR (CDCl$_3$): 2.03 (2H, m, H$^{5'a}$, H$^{5'b}$), 2.58 (2H, m, H$^{4'a}$, H$^{4'b}$), 2.70 (2H, m, H$^{3'a}$, H$^{3'b}$), 3.95 (3H, s, OCH$_3$), 5.97 (1H, t, ^{3}J$_{2'-3'}$ = 2.0, H$^{2'}$), 7.01 (d, 1H, ^{3}J$_{o-m}$ = 8.4, H^o), 7.08 (1H, t, ^{3}J$_{4-5}$ = 7.5, ^{3}J$_{4-3}$ = 7.5, H^4), 7.13 (1H, t, ^{3}J$_{p-m}$ = 7.6, ^{3}J$_{p-m'}$ = 7.6, H^p), 7.20 (1H, dd, ^{3}J$_{3-4}$ = 7.5, ^{4}J$_{3-5}$ = 1.2, H^3), 7.28 (1H, dt, ^{3}J$_{5-6}$ = 8.2, ^{3}J$_{5-4}$ = 7.5, ^{4}J$_{5-3}$ = 1.2, H^5), 7.48 (1H, dt, ^{3}J$_{m-o}$ = 8.4, ^{3}J$_{m-p}$ = 7.6, ^{3}J$_{m-m'}$ = 1.6, H^m), 7.30 (1H, dd, ^{3}J$_{m'-p}$ = 7.6, ^{3}J$_{m'-m}$ = 1.6, H$^{m'}$), 8.44 (1H, d, ^{3}J$_{6-5}$ = 8.2, H^6). ^{13}C NMR (CDCl$_3$): 23.56 (C$^{5'}$), 33.92 (C$^{4'}$), 36.56 (C$^{3'}$), 55.86 (OCH$_3$), 111.40 (C^o), 121.57 (C$^{p'}$), 122.15 (C^6), 127.67 (C^5), 128.0 (C^3), 129.48 (C^1), 130.49 (C^2), 132.71 (C$^{m'}$), 133.14 (C^m), 135.54 (C$^{p'}$), 141.04 (C^2), 157.15 (C^o), 163.26 (C=O).

N-(6-cyclopent-1-en-1-ylcyclohexa-1,5-dien-1-yl)-2-iodobenzamide (16): dark solid (61%). mp. 85–87°C. ^{1}H NMR (CDCl$_3$): 2.03 (2H, m, H$^{5'a}$, H$^{5'b}$), 2.56 (2H, m, H$^{4'a}$, H$^{4'b}$), 2.74 (2H, m, H$^{3'a}$, H$^{3'b}$), 5.98 (1H, t, ^{3}J = 2.0, H$^{2'}$), 7.191 (2H, m, H^4, H^p), 7.28 (1H, dd, ^{3}J$_{3-4}$ = 7.4, ^{4}J$_{3-5}$ = 1.4, H^3), 7.35 (1H, ddd, ^{3}J$_{5-6}$ = 8.2, ^{3}J$_{5-4}$ = 7.4, ^{4}J$_{5-3}$ = 1.4, H^5), 7.47 (1H, t, ^{3}J$_{m-o}$ = 7.4, ^{3}J$_{m-p}$ = 7.4, H^m), 7.52 (1H, d, ^{3}J$_{o-m}$ = 7.4, H^o), 7.95 (1H, d, ^{3}J$_{m'-p}$ = 8.0, H$^{m'}$), 7.97 (1H, br, NH), 8.48 (1H, d, ^{3}J$_{6-5}$ = 8.2, H^6). ^{13}C NMR (CDCl$_3$): 23.34 (C$^{5'}$), 33.90 (C$^{4'}$), 37.11 (C$^{3'}$), 92.29 (C^o), 121.34 (C^6), 124.53 (C^p), 127.80 (C^o), 127.95 (C^5), 128.11 (C^3), 128.45 (C^m), 129.08 (C$^{1'}$), 131.02 (C$^{2'}$), 131.47 (C^4), 134.31 (C), 134.31 (C), 140.23 (C$^{m'}$), 140.51 (C^2), 142.39 (C$^{p'}$), 167.19 (C=O).

N-(6-cyclopent-1-en-1-ylcyclohexa-1,5-dien-1-yl)-3-iodobenzamide (17): dark solid (63%). mp. 86–88°C. ^{1}H NMR (CDCl$_3$): 2.11 (2H, m, H$^{5'a}$, H$^{5'b}$); 2.68 (2H, m, H$^{4'a}$, H$^{4'b}$); 2.77 (2H, m, H$^{3'a}$, H$^{3'b}$); 6.01. (1H, t, ^{3}J = 2.0, H$^{2'}$); 7.17 (1H, dd, ^{3}J$_{3-4}$=7.5, ^{4}J$_{3-5}$=1.2, H^3), 7.24 (1H, dd, ^{3}J$_{m-o}$ = 8.0,

$^3J_{m-p}$ = 7.4, H^m), 7.28 (1H, dd, $^3J_{4-5}$ = 7.5, $^3J_{4-3}$=7.5, H^4), 7.34 (1H, ddd, $^3J_{5-6}$ = 7.5, $^3J_{5-6}$ = 7.5, $^3J_{5-4}$ = 7.5, $^4J_{5-3}$ = 1.2, H^5), 7.82 (1H, d, $^3J_{p-m}$ = 7.4, H^p), 7.91 (1H, d, $^3J_{o-m}$ = 8.0, H^o), 8.21 (1H, s, H$^{o'}$), 8.45 (1H, d, $^3J_{6-5}$ = 7.5, H^6); 8.50 (1H, br, NH). ^{13}C NMR (CDCl$_3$): 23.44 (C$^{5'}$), 29.73 (C$^{4'}$), 33.94 (C$^{3'}$), 94.58 (C$^{m'}$), 120.91 (C^6), 124.32 (C^3), 126.07 (C^p), 127.79 (C^5), 127.92 (C^4), 130.54 (C^m), 130.94 (C$^{2'}$), 134.39 (C^2), 136.21 (C$^{o'}$), 138.97 (C$^{p'}$), 140.59 (C^o), 141.05 (C$^{1'}$), 142.33 (C^1), 169.10 (C=O).

2,4-dichloro-N-(6-cyclopent-1-en-1-ylcyclohexa-1,5-dien-1-yl)benzamide (18): white solid (65%). mp. 78–79°C. ^{1}H NMR (CDCl$_3$): 2.05 (2H, m, H$^{5'a}$, H$^{5'b}$), 2.58 (2H, m, H$^{4'a}$, H$^{4'b}$), 2.72 (2H, m, H$^{3'a}$, H$^{3'b}$), 5.92 (1H, t, J = 18, H$^{2'}$), 7.20 (2H, m, H^4, H^3), 7.35 (1H, dt, $^3J_{5-4}$ = 7.6, $^3J_{5-6}$ = 7.6, H^5), 7.40 (1H, dd, $^3J_{m-o}$ = 8.3, $^4J_{m-m'}$ = 1.4, H^m), 7.51 (1H, d, $^4J_{m'-m}$ = 1.8, H$^{m'}$), 7.77 (1H, d, $^3J_{o-m}$ = 8.3, H^o), 8.46 (1H, br, NH), 8.48 (1H, dd, $^3J_{6-5}$ = 7.6, $^4J_{6-4}$ = 1.0, H^6). ^{13}C NMR (CDCl$_3$): 23.37 (C$^{5'}$), 33.87 (C$^{4'}$), 37.14 (C$^{3'}$), 121.15 (C^6), 124.52 (C^4), 127.79 (C^m), 127.82 (C^5), 128.02 (C^3), 129.21 (C^p), 130.28 (C$^{m'}$), 131.20 (C$^{2'}$), 131.59 (C^o), 133.77 (C$^{o'}$), 134.37 (C^2), 137.15 (C$^{p'}$), 140.43 (C^1), 163.27 (C=O).

N-(2-cyclopent-1-en-1-ylphenyl)pyridine-2-carboxamide (25): yellow solid (79%). mp. 85–87°C. ^{1}H NMR (CDCl$_3$): 2.10 (2H, m, C^{11}H$_2$), 2.65 (2H, m, C^{10}H$_2$), 2.75 (2H, m, C^{9}H$_2$), 6.05 (1H, t, J = 2.2, C^{8}H), 7.20 (1H, t, C^{4}H, J = 7.5), 7.28–7.35 (2H, m, C^{3}H, C^{6}H), 7.48 (1H, t, J = 7.5, C^{5}H), 7.80 (1H, t, J = 6.0, C$^{4'}$H), 8.30 (1H, d, J = 8.0, C^{6}H), 8.05 (1H, d, J = 8.0, C$^{3'}$H), 8.10 (1H, t, J = 8.0, C^{5}H). ^{13}C NMR (CDCl$_3$): 23.51 (C^{10}), 33.91 (C^9), 36.68 (C^{11}), 120.32 (C^6), 122.26 (C$^{3'}$), 123.76 (C^4), 126.13 (C^5), 127.61 (C^8), 128.24 (C^5), 128.93 (C^2), 131.39 (C^3), 134.74 (C^1), 137.48 (C$^{4'}$), 139.94 (C^7), 148.03 (C$^{6'}$), 150.22 (C$^{2'}$), 161.63 (C=O).

12.2.2 GENERAL SYNTHETIC PROCEDURE FOR BENZOXAZINES

To a solution of the amide (0.01 mol) in CH$_2$Cl$_2$ was bubbled HCl$_{(g)}$. After completion of the reaction (control by TLC) was treated 5% NaHCO$_3$, washed with water (1 × 25 mL), dried over MgSO$_4$. The solvent was evaporated, the residue was purified by column chromatography (petroleum ether: ethyl acetate).

2,2'-propane-1,3-diylbisspiro[3,1-benzoxazine-4,1'-cyclopentane] (9): yellow oil (38%). ^{1}H NMR (CDCl$_3$): 2.0 (5H, m, C^2'H, C^4'H, CH$_2$), 2.45 (2H, m, CH$_2$), 2.53 (2H, m, C^5'H), 2.74 (2H, m, C^3'H), 7.04–7.38 (4H, m, Ar). ^{13}C: NMR (CDCl$_3$): 23.54 (C^2, C^5'), 25.35 (CH$_2$), 36.14 (CH$_2$), 40.78 (C$^{3,'}$ C^4'), 88.44 (C^1'), 122.10 (Ar), 123.85 (Ar), 125.91 (Ar), 126.19 (Ar), 127.29 (Ar), 134.90 (Ar), 162.10 (C^1).

2,2'-butane-1,4-diylbisspiro[3,1-benzoxazine-4,1'-cyclopentane] (10): yellow oil (42%). ^{1}H NMR (CDCl$_3$): 1.68 (4H, m, C^2'H, C^5'H), 1.70 (4H, m, C^3'H, C^4'H), 2.10 (2H, m, CH$_2$), 2.21 (2H, m, CH$_2$), 6.95–7.15 (4H, m, Ar). ^{13}C NMR (CDCl$_3$): 23.95(C^2, C^5'), 25.99(CH$_2$), 35.34 (CH$_2$), 40.87 (C$^{3,'}$ C^4'), 88.42 (C^1'), 122.21 (Ar), 124.12 (Ar), 126.22 (Ar), 128.32 (Ar), 128.86 (Ar), 139.01 (Ar), 162.33 (C^1).

2,2'-(1,3-phenylene)bisspiro[3,1-benzoxazine-4,1'-cyclopentane] (11): white solid (82%). mp. 120–122°C. ^{1}H NMR (CDCl$_3$): 1.91 (4H, m, C^{11}H, C$^{11'a}$H), 2.05 (8H, m, C^{12}H, C^{13}H, C$^{13'a}$H, C$^{12'a}$H), 2.35 (4H, m, C^{14}H, C$^{14'a}$H), 7.19–7.30 (8H, m, C^{7}H, C^{7a}H, C^{8}H, C^{8a}H, C^{9}H, C^{9a}H, C^{10}H, C^{10a}H), 7.50 (1H, t, J = 7.8, C^5'H), 8.27 (2H, d, J = 7.8, C^4'H, C^6'H), 8.27 (1H, s, C^2'H). ^{13}C NMR (CDCl$_3$): 23.80 (C^{12}, C^{13}, C^{12a}, C^{13a}), 40.20 (C^{11}, C^{14}, C^{11a}, C^{14a}), 89.16 (C^4, C^{4a}), 122.06 (C^9, C^{9a}), 124.99 (C^2'), 126.59 (C^7, C^{7a}), 127.16 (C^5'), 128.18 (C^{10,} C^{10a}), 128.38 (C^4' C^6'), 129.19 (C^1' C^3'), 130.56 (C^8, C^{8a}), 133.41 (C^5, C^{5a}), 139.59 (C^6, C^{6a}), 156.20 (C^2, C^{2a}).

2,2'-(1,4-phenylene)bisspiro[3,1-benzoxazine-4,1'-cyclopentane] (12): yellow oil (62%). ^{1}H NMR (CDCl$_3$): 1.82 (1H, m, C^2'H), 2.0 (4H, m, C^2'H, C^4'H), 2.25 (2H, m, C^5'H), 7.09–7.25 (6H, m, Ar). ^{13}C NMR (CDCl$_3$): 23.80 (C$^{2,'}$ C^5'), 40.14 (C$^{3,'}$ C^4'), 89.08 (C^1'), 122.04 (Ar), 123.04 (Ar), 126.74 (Ar), 127.59 (Ar), 128.39 (Ar), 129.23 (Ar), 135.62 (Ar), 139.52 (Ar), 156.15 (C^1').

2-(2-nitrophenyl)spiro[3,1-benzoxazine-4,1'-cyclopentane] (19): yellow solid (82%). mp. 79–81°C. ^{1}H NMR (CDCl$_3$): 1.90 (2H, m, H$^{4'a}$, H$^{4'b}$), 2.16 (4H, m, H$^{3'a}$, H$^{3'b}$, H$^{5'a}$, H$^{5'b}$), 2.41 (2H, m, H$^{2'a}$, H$^{2'b}$), 7.18 (1H, d, $^3J_{5-6}$ = 7.5, H^5), 7.20 (1H, dd, $^3J_{6-5}$ = 7.5, $^3J_{6-7}$ = 8.0, H^6), 7.30 (2H, m, H^7, H^8), 7.62 (1H, m, H^m), 7.80 (1H, dt, $^3J_{p-m}$ = 8.0, $^4J_{p-m}$ = 8.0, $^4J_{p-o}$ = 1.7, H^p), 8.27 (1H, dd, $^3J_{m'-p}$ = 8.0, $^3J_{m'-m}$ = 1.7, H-Ar^m), 8.89 (1H, dd, $^3J_{o-m}$ = 7.8, $^4J_{o-p}$ = 1.7, H^o). ^{13}C NMR (CDCl$_3$): 24.0 (C$^{3,'}$ C^4'), 40.65 (C$^{2,'}$ C^5'), 89.80 (C^4), 122.02 (C^5), 123.45 (C^o), 125.32 (C^8), 127.31 (C^6), 128.59 (C^7), 131.52 (C^m), 133.43 (C^p), 134.11 (C$^{m'}$), 135.46 (C$^{p'}$), 138.93 (C^{8a}), 146.85 (C$^{o'}$), 159.15 (C^2).

2-(3-nitrophenyl)spiro[3,1-benzoxazine-4,1'-cyclopentane] (20): brown oil (90%). ^{1}H NMR (CDCl$_3$): 1.92 (2H, m, H$^{4'a}$, H$^{4'b}$), 2.12 (4H, m, H$^{3'a}$, H$^{3'b}$, H$^{5'a}$, H$^{5'b}$), 2.35 (2H, m, H$^{2'a}$, H$^{2'b}$), 7.16 (1H, d, $^3J_{5-6}$ = 7.5, H^5), 7.22 (1H, dd, $^3J_{6-5}$ = 7.5, $^3J_{6-7}$ = 8.0, H^6), 7.31 (2H, m, H^7, H^8), 7.61 (1H, t, $^3J_{m'-o}$ = 8.0, $^3J_{m'-p}$ = 8.0, H-Ar$^{m'}$), 8.32 (1H, dt, $^3J_{o'-m}$ = 8.0, $^4J_{o'-p}$ = 1.7, $^4J_{o'-o}$ = 1.7, H-Ar$^{o'}$), 8.45 (1H, dt, $^3J_{p-m}$ = 8.0, $^4J_{p-o}$ = 1.7, $^4J_{p-o}$ = 1.7, H-Ar^p), 8.92 (1H, dd, $^4J_{o-o'}$ = 1.7, $^4J_{o-p}$ = 1.7, H-Ar^o). ^{13}C NMR (CDCl$_3$): 23.99 (C$^{3'}$, C$^{4'}$), 40.62 (C$^{2'}$, C$^{5'}$), 89.86 (C^4), 122.31 (C^5), 122.71 (C^o), 125.24 (C^8), 127.34 (C^6), 128.64 (C^7), 129.27 (C^{5a}, C$^{m'}$), 133.43 (C^p), 135.16 (C$^{p'}$), 138.93 (C^{8a}), 148.33 (C^m), 154.40 (C^2).

2-(2-methoxyphenyl)spiro[3,1-benzoxazine-4,1'-cyclopentane] (21): brown oil (69%). ^{1}H NMR (CDCl$_3$): 1.85 (2H, m, H$^{5'a}$, H$^{5'b}$), 2.02 (4H, m, H$^{2'a}$, H$^{2'b}$, H$^{4'a}$, H$^{4'b}$), 2.43 (2H, m, H$^{3'a}$, H$^{3'b}$), 3.76 (3H, s, OCH$_3$), 6.94 (1H, d, $^3J_{m'-p}$ = 8.3, H$^{m'}$), 6.99 (1H, t, $^3J_{m-p}$ = 7.5, $^3J_{m-o}$ = 7.5, H^m), 7.12 (1H, d, $^3J_{5-6}$ = 8.3, H^5), 7.18 (1H, m, H^6), 7.27 (1H, m, H^7, H^8), 7.40 (1H, ddd, $^3J_{p-m}$ = 8.3, $^3J_{p-m}$ = 7.5, $^4J_{p-o}$ = 1.7, H^p), 7.65 (1H, dd, $^3J_{o-m}$ = 8.4, $^3J_{m-p}$ = 7.5, $^4J_{o-p}$ = 1.7, H^o). ^{13}C NMR (CDCl$_3$): 24.19 (C$^{2'}$, C$^{5'}$), 40.47 (C$^{3'}$, C$^{4'}$), 55.68 (OCH$_3$), 89.47 (C^4), 120.39 (C^m), 122.06 (C^5), 123.44 (C$^{p'}$), 124.83 (C^7), 128.56 (C^6), 128.28 (C^8), 129.39 (C^{5a}), 131.08 (C^o), 131.85 (C^p), 139.75 (C^{8a}), 158.21 (C$^{o'}$), 158.54 (C^2).

2-(2-iodophenyl)spiro[3,1-benzoxazine-4,1'-cyclopentane] (22): brown oil (81%). ^{1}H NMR (CDCl$_3$): 1.82 (2H, m, H$^{5'a}$, H$^{5'b}$), 2.05 (4H, m, H$^{2'a}$, H$^{2'b}$, H$^{4'a}$, H$^{4'b}$), 2.48 (2H, m, H$^{3'a}$, H$^{3'b}$), 7.02 (1H, t, $^3J_{m-p}$ = 7.5, $^3J_{m-o}$ = 7.5, H^m), 7.14 (1H, d, $^3J_{5-6}$ = 8.3, H^5), 7.19 (1H, m, H^6), 7.29 (1H, m, H^7, H^8), 7.40 (1H, m, H^p), 7.58 (1H, dd, $^3J_{o-m}$ = 8.4, $^3J_{m-p}$ = 7.5, $^4J_{o-p}$ = 1.7, H^o), 7.90 (1H, d, $^3J_{m'-p}$ = 7.9, H$^{m'}$). ^{13}C NMR (CDCl$_3$): 24.19 (C$^{2'}$, C$^{5'}$), 40.47 (C$^{3'}$, C$^{4'}$), 89.47 (C^4), 92.80 (C$^{o'}$), 122.53 (C^5), 124.83 (C^7), 128.28 (C^8), 128.56 (C^6), 129.18 (C^m), 129.42 (C^{5a}), 130.24 (C^o), 131.36 (C^p), 133.76 (C$^{p'}$), 139.10 (C$^{m'}$), 139.45 (C^{8a}), 159.74 (C^2).

2-(3-iodophenyl)spiro[3,1-benzoxazine-4,1'-cyclopentane] (23): brown oil (82%). ^{1}H NMR (CDCl$_3$): 1.59 (2H, m, H$^{5'a}$, H$^{5'b}$), 2.11 (4H, m, H$^{2'a}$, H$^{2'b}$, H$^{4'a}$, H$^{4'b}$), 2.35 (2H, m, H$^{3'a}$, H$^{3'b}$), 7.18–7.26 (3H, m, H^5, H^8, H^m), 7.34 (2H, m, H^6, H^7), 7.86 (1H, ddd, $^3J_{o-m}$ = 6.9, $^3J_{o-o}$ = 1.6, $^4J_{o-p}$ = 1.1, H^o), 8.11 (1H, ddd, $^3J_{p-m}$ = 7.8, $^4J_{p-o}$ = 1.4, H^p), 8.51 (1H, dd, $^4J_{o'-o}$ = 1.6, $^4J_{o'-p}$ = 1.4, H$^{o'}$). ^{13}C NMR (CDCl$_3$): 23.96 (C$^{2'}$, C$^{5'}$), 40.37 (C$^{3'}$, C$^{4'}$), 89.37 (C^4), 93.99 (C$^{m'}$), 122.19 (C^5), 125.03 (C^6), 126.89 (C^7), 126.98 (C^p), 128.52

(C^m), 129.33 (C^{5a}), 129.91 (C^8), 135.29 ($C^{p'}$), 136.66 ($C^{o'}$), 139.37 (C^{8a}), 140.04 (C^o), 155.22 (C^2).

2-(2,4-dichlorophenyl)spiro[3,1-benzoxazine-4,1'-cyclopentane] (24): colorless oil (70%). ^1H NMR ($CDCl_3$): 1.62 (2H, m, $H^{5'a}$, $H^{5'b}$), 2.18 (4H, m, $H^{2'a}$, $H^{2'b}$, $H^{4'a}$, $H^{4'b}$), 2.37 (2H, m, $H^{3'a}$, $H^{3'b}$), 7.22 (1H, dt, $^3J_{4-3}$ = 7.6, $^3J_{4-5}$ = 7.6, $^4J_{4-6}$ = 1.0, H^4), 7.28 (1H, dd, $^3J_{3-4}$ = 7.6, $^3J_{3-5}$ = 1.4, H^3), 7.32 (1H, m, H^5, H^6), 7.44 (1H, dd, $^3J_{m-o}$ = 8.3, $^4J_{m-m'}$ = 1.4, H^m), 7.61 (1H, d, $^3J_{o-m}$ = 8.3, H^o), 7.85 (1H, d, $^4J_{m'-m}$ = 1.8, $H^{m'}$). ^{13}C NMR ($CDCl_3$): 24.24 ($C^{2',}$ $C^{5'}$), 40.56 ($C^{3',}$ $C^{4'}$), 90.12 (C^4), 122.08 (C^5), 125.96 (C^7), 126.71 ($C^{p'}$), 127.14 (C^6), 128.12 (C^m), 129.34 (C^8), 129.49 (C^{5a}), 130.34 (C^o), 131.59 ($C^{m'}$), 133.17 ($C^{o'}$), 136.67 (C^p), 138.56 (C^{8a}), 159.62 (C^2).

2-pyridin-2-ylspiro[3,1-benzoxazine-4,1'-cyclopentane] (26): yellow oil (75%). ^1H NMR ($CDCl_3$): 1.60 (2H, m, $H^{4'}$), 2.0 (4H, m, $H^{2',}$ $H^{5'}$), 2.30 (2H, m, $H^{3'}$), 7.13 (1H, t, J = 7.5, H^6), 7.20 (1H, t, J = 7.5, H^7), 7.32 (2H, m, H^5, H^{11}), 7.47 (1H, d, J = 7.5, H^8), 7.75 (1H, t, J = 7.8, H^{12}), 8.06 (1H, d, J = 7.8, H^{1o}), 8.79 (1H, d, J = 7.8, H^{13}). ^{13}C NMR ($CDCl_3$): 23.59 (C-2,' C-5'), 40.10 (C-3,' C-4'), 89.35 (C-4), 121.94 (Ar), 123.04 (Ar), 125.09 (Ar), 125.67 (Ar), 127.08 (Ar), 128.31 (Ar), 129.42 (Ar), 136.53 (Ar), 139.04 (Ar), 149.66 (Ar), 150.65 (Ar), 155.25 (C=N).

KEYWORDS

- 3,1-benzoxazine
- intramolecular heterocyclization
- ortho-(cyclopent-1-enyl)-aniline
- polyisobutylene

REFERENCES

1. Grovachevskaya E. V., Kvitkovsky F. V., Kosulina T.P. Russian Journal of Chemical Heteroatom Compounds (in Rus.). 2003. №2. 161–320.
2. Kazaryantz S. A., Salikkhov Sh. M., Abdrakhmanov I. B., Ivanova S. R. Bashkirian Chemical Journal (in Rus.). 2009. v.16. №4. 19–24.

3. Kazaryantz S. A., Ivanova S. R. Salikkhov Sh. M., Abdrakhmanov I. B., Islamova R. M. Bashkirian Chemical Journal (in Rus.). 2010. v.17. №3. 42–45.
4. Kazaryantz S. A., Ivanova S. R. Zaripov R. R., Yakupova L. R., Salikhov Sh. M. Herald of Bashkirian University Journal (in Rus.). 2010. v.15. №3. 581–584.
5. Galkin E. G., Erastov A. C., Vyrypaev E. M., Furley I. I Abdrakhmanov I. B Salikhov Sh. M., Krasko S. A. Russian Journal of Chemical Heteroatom Compounds (in Rus.). 2013. №7. 1160–1165.

CHAPTER 13

POLYCONJUGATED LINEAR AND CARDO OLIGOMERS BASED ON AROMATIC DIAMINES AND THEIR THERMAL PROPERTIES

A. F. YARULLIN,[1] L. E. KUZNETSOVA,[1] A. F. YARULLINA,[2] KH. S. ABZALDINOV,[1] O. V. STOYANOV,[1] and G. E. ZAIKOV[1]

[1]Kazan National Research Technological University, Kazan, Russia

[2]Kazan National Research Technical University Named After A.N. Tupolev, Kazan, Russia; E-mail: aleksej-yarullin@yandex.ru, abzaldinov@mail.ru

CONTENTS

ABSTRACT

The series of oligoheteroarilenamines of linear and cardo-structure based on the aromatic diamines were synthesized. The influence of structure on thermal stability in the air was investigated.

13.1 INTRODUCTION

The thermal stability of polymers and oligomers with a system of conjugated double bonds with different heteroatoms or groups in the main chain depends on the strength of bonds in the macromolecule at high temperatures [1–5]. Aromatic oligoheteroarilenamines, relating to this class of compounds, are characterized by the presence of paramagnetic centers, semiconductor temperature dependence of electrical conductivity, superior thermal stability [6, 7]. The availability of their optical properties and enhanced reactivity makes them promising materials in terms of practical use.

13.2 EXPERIMENTAL

Oligoheteroarileamines were synthesized by equilibrium polycondensation reaction in the melt of the initial monomers [8]. Linear oligoarilenamines based on the hydroquinone and some aromatic diamines and cardo materials, based on the same diamines with phenolphthalein were synthesized by the same method. P-phenylenediamine (p-PDA), 4,4'-diaminediphenylmethane (DADPM), 4,4'-diaminediphenylsulfone (DADPS), 4,4'-diaminediphenylsulfide (DADPSd), 4.4'diaminediphenylether (DADPE), 4,4'-diaminediphenyl (DADP), 2,4-diamine-6-phenyl-1,3,5-triazine (DAPTr) were used as diamines.

The studying of the obtained oligoarilenamines oxidative degradation process was carried out by thermogravimetric analysis (Paulik-Paulik-Erdey derivatograph) (TGA) in the temperature range 20–600°C, at a heating rate of 6 deg./min, at the dynamic mode, in air.

13.3 RESULTS AND DISCUSSION

The purpose of this study was to investigate the influence of structure of the synthesized cardo (based on phenolphthalein) (Table 13.1) and linear (the hydroquinone) (Table 13.2) oligoheteroarilenamines, that contain different heteroatoms or groups of atoms in the conjugated chain, with different bond dissociation energies on their thermal stability.

The initial period of Oligoheteroarilenamines thermal degradation (Figs 13.1 and 13.2), corresponding to an insignificant mass loss, due to the fact that the investigated powder samples easily absorb the air humidity and in the initial period of heating the appearance of small endothermic peaks on DTA curves is noted. Moreover, these peaks may be associated with the solid-state polycondensation (SSP) process and isolation of the low molecular weight reaction product. The polarized relations and multicenters in the structure of oligoheteroarilenamines macromolecules provide chemical strength of these bonds and hence the increased thermal stability of the whole system.

TABLE 13.1 The Comparative Evaluation of Thermal Stability of Oligoheteroarilenamines Based on Phenolphthalein and Several Diamines [9]

№	Oligoheteroarilenamines	$T_{(\Delta m=5\%)}$, °C	$T_{(\Delta m=10\%)}$, °C	$T_{(\Delta m=50\%)}$, °C
1	2	3	4	5
1		300	330	525
2		300	330	495
3		310	320	495
4		300	335	420
5		275	290	525
6		270	300	510
7		280	310	500

TABLE 13.2 The Comparative Evaluation of Thermal Stability of Oligoheteroarilenamines Based on Hydroquinone and Several Diamines

№	Oligoheteroarilenamines	$T_{(\Delta m=5\%)}$, °C	$T_{(\Delta m=10\%)}$, °C	$T_{(\Delta m=50\%)}$, °C
1		290	315	500
2		275	300	370
3		325	340	380
4		370	385	450
5		365	380	470
6		375	425	470
7		270	300	370

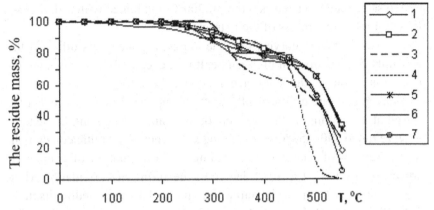

FIGURE 13.1 TGA–curves of oligoamines based on phenolphthalein (PP) and the number of diamines: 1 – p-PDA; 2 – DADPM; 3 – DADPS; 4 – DADPSd; 5 – DADPE; 6 – DADP; 7 – DAPTr.

FIGURE 13.2 TGA–curves of oligoamines based on hydroquinone (Hq) and the number of diamines: 1 – p-PDA; 2 – DADPM; 3 – DADPS; 4 – DADPSd; 5 – DADPE; 6 – DADP; 7 – DAPTr.

Oligoheteroarilenamines destruction during thermal exposure begins with breaking of the weakest links, especially in places of disturbed pairing. Besides that, the destruction can be accompanied by the destruction of phthalide cycle in cardo oligoamines. It was noticed that, during comparing of thermal stabilities of oligoheteroarilenamines based on the hydroquinone and phtalide containing materials, the nature of the TGA curve has the same angle to a temperature of 200°C and linked with a slight loss of the samples, regardless of their structure.

As seen from the curves, phtalide containing materials differ in the proximity of placement at high temperatures, except oligoamine, containing sulfone group in the diamine component (DADPS)

At a temperature of 300°C all tested oligoamines lose between 8–12% of weight, depending on the structure of the diamine fragment. In the temperature range 300–350°C crosslinking and structure simultaneously with the destruction of chains occur, leading to the formation of condensed structures. In view of a small degree of the sulfur atom oxidation while heating of oligoheteroarylenamine based on 4,4'-diaminedifenilsulfide (DADPSd) (Fig. 13.1: curve 4, Fig. 13.2: curve 1), the interaction of it with oxygen happens with forming of a sulfoxide, which is able to further oxidation to the hexavalent sulfur with sulfide forming. During heating of oligoamine based on 4,4'-diaminediphenylsulfone (DADPS) (Fig. 13.1: curve 3, Fig. 13.2: curve 4), sulfone formation and SO_2 selection may occur as a result of degradation of the carbon – sulfur bond. Therefore the curve of mass loss of oligoarilene based on 4,4'-diaminediphenylsulfone is shifted toward large losses with respect to the weight loss curve of oligoarilene based on 4,4'-diaminedifenilsulfide and the percentage of weight loss at the same temperature will be higher. Thus at 350°C the losses are between 10% and 40%, respectively.

The oligoamines, which are containing diphenylene units (DADP) (Fig. 13.1: curve 6, Fig. 13.2: curve 3), are the most resistant to degradation in air, their weight loss at 400°C is less than 5% in the case of a linear oligomer.

The sharp mass decline of all oligoheteroarilenamines samples, which is associated with their deep degradation, starts at temperatures above 400°C and is linked to rupture of the carbon – nitrogen bond in the main chain. In this case, the slope of the curves of thermal destruction is almost the same for all. The oligoamines thermal destruction leads to a sharp decrease in the flexibility of the macromolecular chains as a result of its collapse into shorter sections, as well as the flow of structuring processes.

When heating of polyconjugated oligoamines, EPR signal change (increase in the intensity and width reduction) is observed, which is symbatic to its weight loss change. This is due to the fact that in the process of destruction self-stabilization the available oligomer paramagnetic particles are involved.

It is shown that a linear oligomers, based on the hydroquinone, possess thermal stability, they lose at a temperature of 300°C an average of 15% by weight and at a temperature of 400°C their temperature resistance decreases in the series:

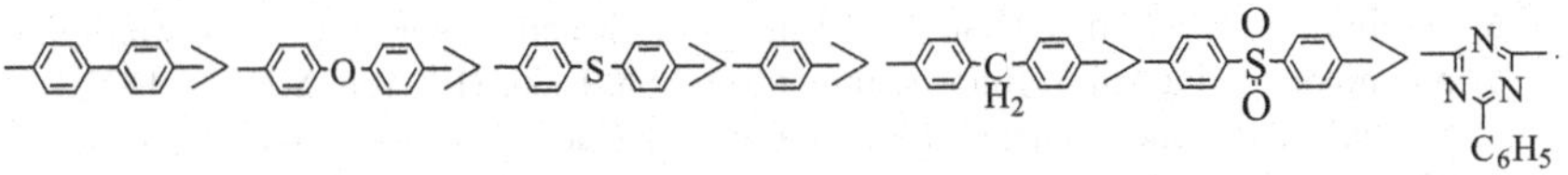

For cardo oligoamines based on phenolphthalein the weight loss in these conditions make up about 10% by weight, which is associated with the fact that phthalide groups contribute to the thermal stability of the system. So, 50% weight loss for them are in high temperature area from 420°C to 525°C.

The chemical structure of oligoheteroarilenamines diamine fragment primarily determines their thermal stability, which should be considered when determining the reaction conditions for their preparation, as the partial destruction can happen during oligomer synthesis.

KEYWORDS

- cardo structure
- linear structure
- oligoheteroarilenamine
- thermal stability
- weight loss

REFERENCES

1. Berlin A. A. O nekotorych problemach khimii polymerov s sistemoy sopriagenia. A.A. Berlin. Chimicheskaya promyshlennost. 1992. No 12. P. 25. (in Russian).
2. Vinogradova S.V. Polykondensatsionnye processy i polymery. S.V. Vinogradova, V.A. Vasnev. Moscow: "Nauka," 2000. 373 pp. (in Russian).
3. K. Miyatake, B. Bae, M. Watanabe Fluorene-containing cardo polymers as ion conductive membranes for fuel cells. K. Miyatake, B. Bae, M. Watanabe. Polymer Chemistry, 2011, Vol. 2, P. 1919–1929.

4. A. Ravve Principles of polymer chemistry. A. Ravve. Springer Science & Business Media, 2000. Vol. 1. 626 pp.

5. S. N. Salazkin, V.V. Shaposhnikova Synthesis and behavior of phthalide-containing polymers. S.N. Salazkin, V.V. Shaposhnikova. Polymers for Advanced Technologies. G.E. Zaikov, L.I. Bazyljak, J.N. Aneli (editors). Apple Academic Press, Inc. 2013. P. 155–174.

6. G. Sh. Liou, H.J. Yen, Y.T. Su, H.Y. Lin Synthesis and properties of wholly aromatic polymers bearing cardo fluorine moieties. G.Sh. Liou, H.J. Yen, Y.T. Su, H.Y. Lin. Journal of Polymer Science. Part A: Polymer Chemistry. 2007. Vol. 45. P. 4352–4363.

7. P. E. Cassidy Thermally Stable Polymers: Synthesis and Properties. P. E. Cassidy. Marcel Dekker: New York, 1980; p 179.

8. Yarullin A.F. Izuchenie kinetiki processa polucheniya oligoaminov i oligoamidov metodom ravnovesnoy polycondensatsii. A.F. Yarullin, L.E. Kuznetsova, L.F. Zakirova. The articles of XV All-Russian Conference "Structura i dynamica molekuliarnikh sistem." Yalchik, Russia, 2008. p. 297 (in Russian).

9. Yarullin A.F. Electrophysical Properties of Oligomer–Polymer Complexes Based on Heat_Resistant Oligoaryleneamines. L.E. Kuznetsova, A.F. Yarullina, O.V. Stoyanov. MAIK "Nauka. Interperiodica," Polymer Science, Series D. Glues and Sealing Materials, 2013. Vol. 6, No. 2, 109–115.

CHAPTER 14

ALKYLATION OF AROMATIC HYDROCARBONS BY POLYHALOGEN-CONTAINING CYCLOPROPANES

A. N. KAZAKOVA,[1] G. Z. RASKILDINA,[1] N. N. MIKHAYLOVA,[1] L. V. SPIRIKHIN,[2] and S. S. ZLOTSKY[1]

[1]*Ufa State Petroleum Technological University, Kosmonavtov Str. 1, 450062 Ufa, Russia; Tel: +(347) 2420854, E-mail: nocturne@mail.ru*

[2]*Institute of Organic Chemistry, Ufa Scientific Center, Russian Academy of Sciences, Oktyabrya avenue 71, 450054 Ufa, Russia*

CONTENTS

ABSTRACT

This chapter presents the results on alkylation of benzene and toluene with 2-chloromethyl-*gem*-dichlorocyclopropane **Ia** and 2-(chloromethyl)-2-methyl-*gem*-dichlorocyclopropane **Ib** in the presence of an aluminum chloride catalyst. Additionally we investigated a reaction of substituted benzenes with 2-bromo-2-phenyl-*gem*-dichlorocyclopropane.

14.1 RESULTS AND DISCUSSION

The data obtained in this study (Table 14.1) show that in the case of alkylation of benzene and toluene with reagents **Ia** and **Ib** in the presence of catalytic amounts of $AlCl_3$, the reaction is accompanied by opening of the cyclopropane ring and leads to the formation of the corresponding 3,3-dichloroalkenylderivatives **IIa, IIb,** and **IIIa–IIId** (Scheme 1).

TABLE 14.1   Alkylation of Benzene and Toluene by *gem*-dichlorocyclopropanes **Ia** and **Ib**

Reactants	Time, h	Products	Yield, %
Benzene + **Ia**	2	**IIa**	75
Benzene + **Ib**	2	**IIb**	21
Toluene + **Ia**	3	**IIIa,b**	70*
Toluene + **Ib**	3	**IIIc,d**	43**

*Mixture of o- and p-isomers (**IIIa:IIIb** = 1:7.5).

** Mixture of o- and p-isomers (**IIIc:IIId** = 1:9).

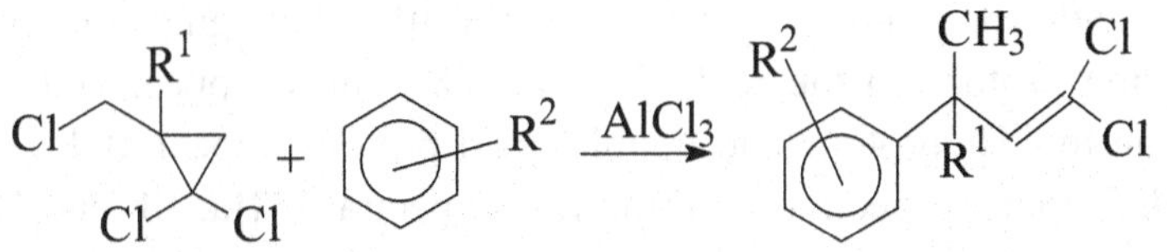

Ia,b   IIa,b, IIIa-d

R^1=H (Ia); R^1=CH_3 (Ib);

R^1=H, R^2=H (IIa); R^1=H, R^2=CH_3 (IIb);

R^1=H, R^2=ortho-CH_3 (IIIa); R^1=H, R^2=ortho-CH_3 (IIIc);

R^1=H, R^2=para-CH_3 (IIIb); R^1=H, R^2=para-CH_3 (IIId).

SCHEME 1

In the case of chloromethyl derivative Ia, the yields of alkylation products are considerably higher than those with the use of chloride IIb. At the same time, in the reaction of Ia and IIb with toluene (Table 14.1) yields the o- and p-isomers in approximately the same ratio (**IIIa:IIIb** = 1:7.5, **IIIc:IIId** = 1:9).

The formation of IIa, IIb, and IIIa–IIId is presumably due to the elimination of HCl molecule, involving cyclopropane ring opening by the C1–C2 bond and the formation of intermediate unstable diene which reacts with the aromatic substrate (Scheme 2).

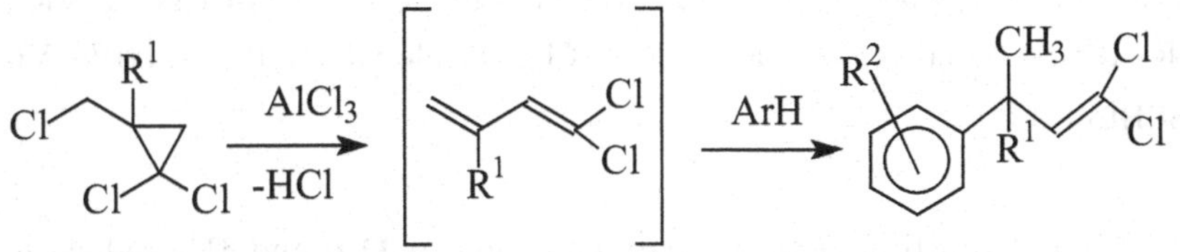

SCHEME 2

This type of ring opening in substituted gem-dichlorocyclopropanes mediated by acid catalysts was described earlier in Ref. [8].

The relative activity of Ia and Ib toward benzene was estimated by means of competitive kinetics and it was found that the presence of methyl group in the cyclopropane moiety reduces the activity of compound Ib by a factor of 3, which is in a good agreement with our data (Table 14.1).

In ^{1}H NMR spectrum of IIa, IIb, and IIIa–IIId, the signals of cyclopropane ring protons in the region of 1.2–1.8 ppm disappear and new signals due to CH_3 protons and double bond protons appear at 1.4–1.5 and 5.9–6.5 ppm, respectively. In the mass spectra of IIa, IIb and IIIc, IIId, the molecular ions have a low intensity to make 2 to 10%, whereas compounds IIIa and IIIb form stable molecular ions (m/z = 28–31%). The most abundant ions are those with m/z = 129 for IIa, m/z = 143 for IIb, IIIa, IIIb, and m/z= 157 for IIIc, IIId formed as a result of successive elimination of halogens. The main route of dissociation is almost the same for the o-and p-isomers of IIIa–IIId.

In the presence of the Friedel–Crafts reaction catalysts ($AlCl_3$) the alkylation of aromatic compounds IVa–IVd with the bromo derivative V occurs through the cyclopropane ring opening to give the corresponding 2-aryl-3-phenyl-1,1-dichlorprop-1-enes VIa–VIf (Scheme 3).

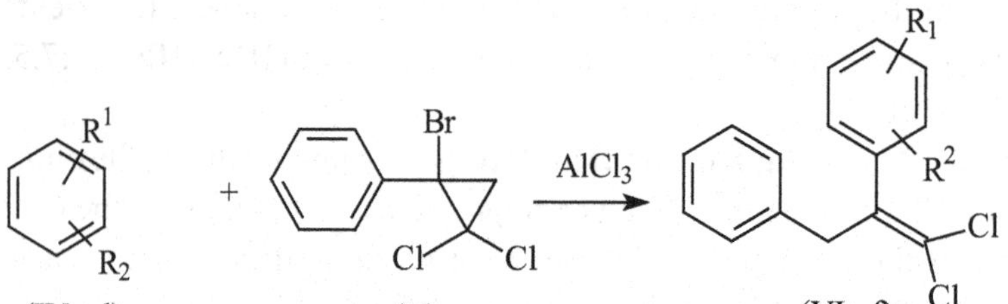

$R^1=R^2=H$ (IVa, VIa); $R^1=H$, $R^2=2$-CH_3 (IVb, VIb); $R^1=H$, $R^2=4$-CH_3 (IVc, VIc); $R^1=R^2=2,4$-CH_3 (IVc, VId); $R^1=H$, $R^2=2$-Cl (IVd, VIe); $R^1=H$, $R^2=4$-Cl (IVd, VIf).

SCHEME 3

Methyl substituents in the aromatic ring of IVb and IVc reduce the activity of an aromatic compound; while the electronegative chlorine in Id has virtually no effect on the yield of alkylation products (Table 14.2).

The formation of products VIa–VIf is probably due to the primary elimination of HBr molecule followed by alkylation of aromatic ring of IVa–IVd with unstable cyclopropene that is accompanied by the ring opening and the migration of a hydrogen atom (Scheme 4).

Earlier the rearrangement of the gem-dichlorocyclopropanes with the formation of 1,1-dichloroalk-1-enes has been described in Refs. [9, 10].

TABLE 14.2 Alkylation of Aromatic Hydrocarbons with 2-bromo-2-phenyl-*gem*-dichlorocyclopropane [(**IVa-IVd**):II:AlCl$_3$ = 7.8:1:0.3, *T*=100°C]

Comp. no.	Reaction products	Time, h	Yield, % (*ortho:para*)
IVa	VIa	2	47
		0.5^a	98
IVb	VIb + VIc	2	37 (1:1)
		0.5^a	65 (2.5:1)
IVc	VId	2	29
		0.5^a	60
IVd	VIe + VIf	2	48 (1:1)
		0.5^a	97 (2:1)

^{a} Microwave irradiation, 420 W.

Note that the microwave irradiation reduces the reaction time and increases almost two times the yield of alkylation products **VIa–VIf**. The selectivity of oisomers formation increases in the case of compounds **IVb** and **IVc** (Table 2).

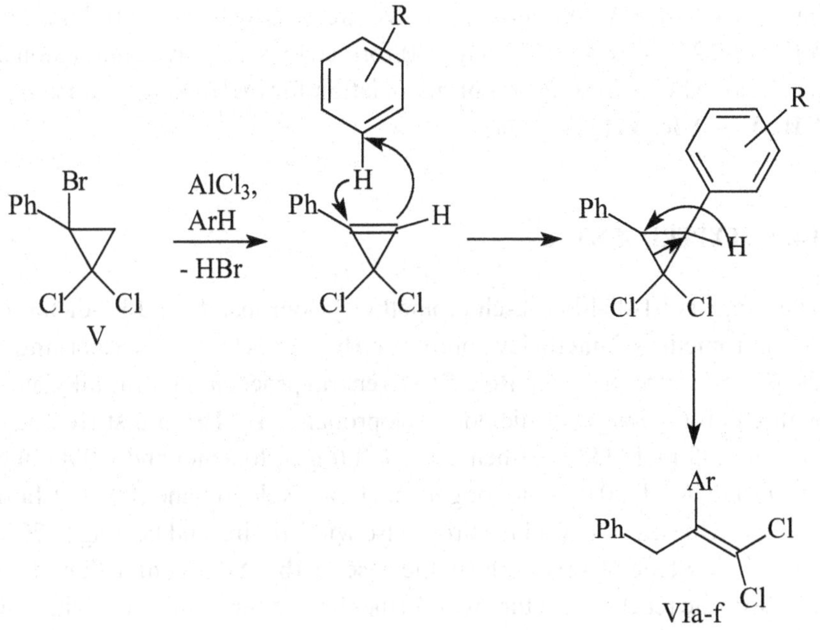

SCHEME 4

The products of the reaction of aromatic compounds **IVa–IVd** with reagent **V** were isolated by vacuum distillation as individual compounds **IIIa, IIId** or as mixtures of isomers **VIb + VIc, VIe + VIf**. Their structure and the ratio of *ortho-* and *para*-isomers in the mixture were determined by the ^{1}H and ^{13}C NMR spectroscopy, GC-MS method.

A feature of the ^{13}C NMR spectra of compounds **VIa–VIf**, recorded in the modulation of the constant of C–H coupling made (JMOD), is the presence of a signal characteristic of the methylene carbon atoms (42.20 **VIa**, 41.89 **VIb**, 39.52 **VIc**, 39.41 **VId**, 41.76 **VIe**, 39.61 ppm **VIf**) and a signal of quaternary carbon atom of CCl2-group (118.83 **VIa**, 118.76 **VIb**, **VIc**, 118.82 **VId**, 119.93 **VIe**, 119.27 ppm **VIf**). The ^{13}C NMR spectra of *para*-isomers contain the mutually equivalent signals of the carbon atoms of double intensity, whereas in the spectra of *ortho*-isomers with different substituents all the carbon atoms of the aromatic rings have different chemical shifts [11]. On this basis the signals of the aromatic carbons in the spectrum of mixtures **VIb + VIc** and **VIe + VIf** were assigned to the corresponding *ortho-* and *para*-isomers.

In the mass spectra of compounds **VIa–VIf** there are stable molecular ions (10–40%). The most intensive are ions with m/z 191 **VIa**, 205 **VIb, VIc**, 219 **VId**, 225/227 **VId, VIf** formed by successive elimination of halogens. The main direction of dissociation for *ortho-* and *para*-isomers **VIb, VIc, VIe, VIf** is similar.

14.2 EXPERIMENTAL

The reagents 1,1-dichloro-2-chloromethylcyclopropan **Ia** and 1,1-dichloro-2-(chloromethyl)-2-methylcyclopropane **Ib** were synthesized according to the procedure described in Ref. [7]. A general procedure for the alkylation with chloromethyl_*gem*_dichlorocyclopropanes **Ia, Ib**: to a stirred mixture of 0.05 mol (3.9 g) of benzene (or 4.6 g of toluene) and 0.0007 mol (0.09 g) of AlCl3, 0.005 mol of *gem*_dichlorocyclopropane (0.8 g of **Ia** or 0.87 g of **Ib**) was slowly added drop-wise with stirring and heating to 70°C for 2 (in the case of **Ia**) or 3 h (in the case of **Ib**). After completion of the addition, the reaction mixture was further heated for 15 min with vigorous stirring and then poured into a beaker with ice and 10% hydrochloric acid and extracted with diethyl ether. The organic layer was washed with water and dried over anhydrous $MgSO_4$. After solvent evaporation, the resulting residue was distilled in vacuum.

The reaction products were analyzed on an LKhM-8 MD chromatograph equipped with a thermal conductivity detector and 2-m column with 5% SE-30 on a Chromatin N-AW support. The starting column temperature was 100°C, and the final one was 300°C, the programming rate was 20°C/min, the carrier gas (helium) flow rate was 30 mL/min, and the evaporator temperature was 250°C. The calculations were performed using the method of internal normalization, which is based on the reduction of a total area of peaks to 100%.

^1H and ^{13}C NMR spectra were recorded on a Bruker AM-300 spectrometer (300.13 and 75.47 MHz, respectively) in $CDCl_3$, using Me_4Si as an internal standard. Mass chromatograms were recorded on a SHIMADZU GCMS-QP2010 Plus instrument (EI, 70 eV, ion source temperature of 200°C, on-column injection temperature of 40–290°C, heating rate of 12°C/min).

Individual substances **IIa** and **IIb** and a mixture of **IIIa, b** and **IIIc, d** isomers were isolated by vacuum distillation from the reaction mass of the interaction of the aromatic compounds with reagents **Ia** and **Ib**. Their structure and the *o*- to *p*-isomer ratio in the mixtures of **IIIa, b** and **IIIc, d** were determined by ^1H NMR and mass spectrometry.

(3,3-Dichloro-1-methylpropene-2-yl-1)benzene (IIa). Yield 75%, colorless liquid. Bp 96°C (5 mmHg). ^1H NMR ($CDCl_3$, δ, ppm): 1.36 (d, 3H, CH_3, ^3J 7.2 Hz), 3.75–3.86 (qu.d, 1H, CH_3–CH, ^3J 7.2 Hz, ^3J 9.6 Hz), 5.93–5.96 (d, 1H, =CH, ^3J 9.6 Hz), 7.11–7.29 (m, 5H, Ar). ^{13}C NMR ($CDCl_3$, δ, ppm): 20.74 (CH_3), 40.31 (CH_3–CH), 120.03 (CCl_2), 126.86 (*m*-Ar), 126.93 (*o*-Ar), 128.84 (*p*-Ar), 134.41 (=CH), 143.54 (Ph). Mass spectrum *m/e*, (I_{rel}, %) 200/202/204 M$^+$ (10/6/L.3), 185/187/18 [M–CH_3]$^+$ (5/3/0.8), 165/167 [M–Cl]$^+$ (18/6) 149/145 (60/20), 129 (100), 115 (23), 105 (12), 91 (4), 77 (30), 63 (20), 51 (37).

Mixture of 1-(3,3-dichloro-1-methylpropene-2-yl-1)-2-methylbenzene (IIIa) and 1-(3,3-dichloro-1-methylpropene-2-yl-1)-4-methylbenzene (IIIb) (ratio of isomers IIIa: IIIb = 1: 7.5). Yield 70%, colorless liquid. Bp 112–114°C (5 mm Hg). *o*-**Isomer (IIIa)**. ^1H NMR ($CDCl_3$, δ, ppm): 1.40 (d, 3H, CH_3, ^3J 6.9 Hz), 2.38 (s, 3H, CH_3–Ar), 3.97–4.07 (qu.d,

1H, CH_3–CH, ^3J 6.9 Hz, ^3J 9.3 Hz), 5.95–5.98 (d., 1H, =CH, ^3J 9.3 Hz), 7.10–7.21 (m, 4H, Ar). ^{13}C NMR (CDCl3, δ, ppm): 20.34 (CH_3–Ar), 21.59 (CH_3), 31.98 (CH_3–CH), 123.15 (CCl_2), 123.89 (*p*-Ar), 125.60, 127.55 (*m*-Ar), 130.64 (*o*-Ar), 135.02 (*o*-Ar–CH_3), 137.02 (=CH), 139.42 (Ar). Mass spectrum *m/e*, (I_{rel}, %): 214/216/218 M⁺ (28/18/3), 199/201/203 [M–C]⁺ (30/20/3.5), 179/181 [M–Cl⁻]⁺ (25/8), 163/165 (98/30), 143 (100), 128 (96), 119 (11), 115 (40), 91 (35), 77 (25), 63 (28), 51 (34).

p-**Isomer (IIIa).** ^1H NMR (CDCl3, δ, ppm): 1.41 (d, 3H, CH_3, ^3J 7.2 Hz), 2.35 (s, 3H, CH_3–Ar), 3.77–3.88 (qu.d, 1H, CH_3–CH, ^3J 7.2 Hz, ^3J 9.6 Hz), 5.97–6.0 (d, 1H, =CH, ^3J 9.6 Hz), 7.10–7.21 (m, 4H, Ar). ^{13}C NMR (CDCl3, δ, ppm): 20.73 (CH_3–Ar), 21.10 (CH_3), 39.86 (CH_3–CH), 123.15 (CCl_2), 126.75 (*m*-Ar), 129.45 (*o*-Ar), 134.55 (=CH), 136.20 (*p*-Ar), 140.54 (Ar). Mass spectrum *m/e*, (I_{rel}, %): 214/216/218 M⁺ (31/21/3), 199/201/203 [M–C]⁺ (40/25/4.5), 179/181 [M–Cl⁻]⁺ (35/12), 163/165 (95/30), 143 (100), 128 (97), 119 (18), 115 (28), 91 (30), 77 (23), 63 (22), 51 (28).

(3,3-Dichloro-1,1-dimethylpropene-2-yl-1)benzene (IIa). Yield 21%, colorless liquid. Bp 109°C (5 mmHg). ^1H NMR (CDCl₃, δ, ppm): 1.55 (s, 6H, CH_3–C–CH_3), 6.25 (s, 1H, =CH), 7.18–7.33 (m, 5H, Ar). ^{13}C NMR (CDCl₃, δ, ppm): 29.53 (CH_3–C–CH_3), 54.66 (CH_3–C–CH_3), 126.25 (*p*-Ar), 128.36 (*o*-Ar), 128.51 (*m*-Ar), 134.26 (CCl_2), 138.57 (=CH), 140.85 (Ar). Mass spectrum *m/e*, (I_{rel}, %): 214/216/218 M⁺ (2), 199/201/203 [M⁻]⁺ (1), 179/181 [M–Cl⁻]⁺ (45/15), 163/165 (18/6), 143 (100), 128 (68), 115 (11), 91 (15), 77 (17), 63 (12), 51 (20).

Mixture of 1-(3,3-dichloro-1,1-dimethylpropene-2-yl-1)-2-methylbenzene (IIIc) and 1-(3,3-dichloro-1,1-dimethylpropene-2-yl-1)-4-methylbenzene (IIId) (ratio of isomers IIIc: IIId = 1: 9). Yield 43%, colorless liquid. Bp 122–124°C (5 mmHg).

o-**Isomer (IIIc).** ^1H NMR (CDCl₃, δ, ppm): 1.54 (s, 6H, CH3–C–CH3), 2.37 (s, 3H, CH_3-Ar), 6.23(s, 1H, CH=), 7.01–7.26 (m, 4H, Ar). ^{13}C NMR (CDCl₃, δ, ppm): 21.14 (CH_3–Ar), 29.60 (CH_3–C–CH_3), 47.02 (CH_3–C–CH_3), 123.07 (*o*-Ar), 125.43 (*p*-Ar), 126.49, 128.28 (*m*-Ar), 134.43 (CCl_2), 135.73 (*o*-Ar–CH_3), 142.90 (=CH), 145.52 (Ar). Mass spectrum *m/e*, (I_{rel}, %): 228/230/232 M⁺ (7/5/0.8), 213/215/217 [M⁻]⁺ (1), 193/196 [M–Cl⁻]⁺ (50/18), 177/179 (22/7), 157 (100), 142 (60), 128 (20), 115 (27), 105 (11), 91 (18), 77 (17), 65 (18), 51 (12).

p-**Isomer (IIId).** ^1H NMR (CDCl₃, δ, ppm): 1.53 (s, 6H, CH_3–C–CH_3), 2.35 (s., 3H, CH_3–Ar), 6.23 (s, 1H, CH=), 7.01–7.26 (m, 4H, Ar). ^{13}C NMR

(CDCl$_3$, δ, ppm): 21.04 (CH$_3$–Ar), 29.60 (CH$_3$–C–CH$_3$), 49.13 (CH$_3$–C–CH$_3$), 125.86 (o-Ar), 129.07 (m-Ar), 134.43 (CCl$_2$), 135.42 (p-Ar–CH$_3$), 138.75 (=CH), 144.71 (Ar). Mass spectrum m/e, (I_{rel}, %): 228/230/232 M$^+$ (8/5/0.8), 213/215/217 [M$^-$]$^+$ (1), 193/196 [M–Cl·]$^+$ (55/18), 177/179 (29/10), 157 (100), 142 (64), 128 (23), 115 (28), 105 (11), 91 (18), 77 (18), 65 (17), 51 (12).

2-Bromo-2-phenyl-*gem*-dichlorocyclopropane (V). Yield 95%, colorless liquid, bp 125–127°C (4 mm Hg). 1H NMR spectrum, δ, ppm (J, Hz): 2.06 d (1H, C^{3}H$_a$, ^2J 9.0), 2.09 d (1H, C^{3}H$_b$, ^2J 9.0), 7.17–7.39 m (4H, Ph). ^{13}C NMR spectrum, δ$_C$, ppm: 35.43 (C^{3}H$_2$), 43.01 (C^{2}Br), 62.88 (C^{1}Cl$_2$), 128.71 (C^{2}H, C^{6}H, Ph), 129.01 (C^{4}H, Ph), 129.28 (C^{3}H, C^{5}H, Ph), 138.87 (C^1). Mass spectrum, m/z (I_{rel}, %): 264/266/268/270 (1) [M]$^+$, 192/194 (5/5), 185/186/188 (10/6/1) [M – Br]$^+$, 149/151 (100/30), 115 (47), 89 (28), 75 (18), 63 (22).

Alkylation of aromatic hydrocarbons IVa–IVd with 2-bromo-2-phenyl-*gem*-dichlorocyclopropane (V) (*general procedure*). To a mixture of 39 mmol of arene **IVa–IVd** and 1.4 mmol of AlCl$_3$ under stirring and heating to 90°C was added drop-wise a solution of 5 mmol of 2-bromo-2-phenyl-*gem*-dichlorocyclopropane **V** in 11 mmol of the corresponding aromatic compound over 2 h. After the addition was completed, the reaction mixture was heated with vigorous stirring for further 15 min and then poured into a mixture of ice and 10% hydrochloric acid solution, then extracted with diethyl ether. The organic layer was washed with water and dried over MgSO$_4$. After evaporation of the solvent the residue was distilled in vacuum.

Alkylation of aromatic hydrocarbons IVa–IVd with 2-bromo-2-phenyl-*gem*-dichlorocyclopropane V under microwave irradiation (*general procedure*). A mixture of 50 mmol of arene **IVa–IVd**, 1.4 mmol of AlCl$_3$ and 5 mmol of 2-bromo-2-phenyl-*gem*-dichlorocyclopropane **V** was stirred under irradiation in a domestic microwave oven Sanyo EM-S1073W at 420 W for 30 min. Then the reaction mixture was poured into a mixture of ice and 10% hydrochloric acid solution and extracted with diethyl ether. The organic layer was washed with water and dried over MgSO$_4$. After evaporation of the solvent the residue was distilled in a vacuum.

1,1'-(1,1-Dichloroprop-1-ene-2,3-diyl)biphenyl (VIa). Colorless liquid, bp 142°C (2 mm Hg). ^{1}H NMR spectrum, δ, ppm: 3.95 s (2H,

C^3H_2), 7.05–7.30 m (10H, Ph). ^{13}C NMR spectrum, δ_C, ppm: 42.30 (C^3H_2), 118.83 (C^1Cl_2), 126.58 (C^4H, Ph), 127.73 ($C^4'H$, Ph), 128.18 (C^3H, C^5H, Ph), 128.38, ($C^3'H$, $C^5'H$, Ph), 128.27 ($C^2'H$, $C^6'H$, Ph), 128.67 (C^2H, C^6H, Ph), 136.99 ($C^{1'}$ Ph), 138.63 (C^1, Ph), 139.05 (C^2). Mass spectrum, m/z (I_{rel}, %): 262/264/266 (16/10/2) $[M]^+$, 227/229 (24/7) $[M - Cl]^+$, 191 (74), 165 (17), 149/151 (100/30), 136 (14), 115 (12), 91 (74), 65 (47), 51 (25).

A mixture of 1-(1-benzyl-2,2-dichlorovinyl)-2-methylbenzene (VIb) and 1-(1-benzyl-2,2-dichlorovinyl)-4-methylbenzene (VIb). Colorless liquid, bp 152–154°C (2 mm Hg). *ortho*-VIb. ^{13}C NMR spectrum, δ_C, ppm: 19.48 (CH_3), 41.89 (C^1H_2, Bn), 118.76 (C^2Cl_2), 125.65 (C^4H, Bn), 130.09 (C^4H, C^6H_4), 126.58 (C^3H, C^5H, Bn), 128.53 (C^3H, C^6H4), 128.41 (C^5H, C^6H_4), 129.07 (C^2H, C^6H, Bn), 125.92 (C^6H, C^6H_4), 136.86 (C^2, C^6H_4), 135.09 (C^1, C^6H_4), 136.02 (C^1, Bn), 138.75 (C^1). Mass spectrum, m/z (I_{rel}, %): 276/278/281 (16/7/4) $[M]^+$, 241/243 (54/16) $[M –Cl]^+$, 205 (100), 191 (18), 178 (12), 163/165 (41/11), 149/151 (54/21), 136 (13), 128 (13), 105 (56), 101 (33), 89 (16), 77 (31), 65 (20), 63 (21), 53 (11), 51 (28). *para*-VIc. ^{13}C NMR spectrum, δ_C, ppm: 20.97 (CH_3), 39.52 (C^1H_2, Bn), 118.76 (C^2Cl_2), 125.65 (C^4H, Bn), 126.58 (C^3H, C^5H, Bn), 128.08 (C^3H, C^5H, C^6H_4), 127.66 (C^2H, C^6H, C^6H_4), 128.29 (C^2H, C^6H, Bn), 137.54 (C^4, C^6H_4), 133.86 (C^1, C^6H_4), 136.38 (C^1, Bn), 140.06 (C^1). Mass spectrum, m/z (I_{rel}, %): 276/278/281 (22/12/2) $[M]^+$, 241/243 (59/12) $[M - Cl]^+$, 205 (83), 191 (12), 189 (17), 178 (12), 163/165 (66/31), 149/151 (41/17), 136 (21), 127 (15), 105 (100), 101 (36), 89 (17), 77 (39), 65 (19), 63 (21), 53 (9), 51 (33).

2-(1-Benzyl-2,2-dichlorovinyl)-1,4-dimethylbenzene(VId). Colorless liquid, bp 163°C (2 mm Hg). ^{13}C NMR spectrum, δ_C, ppm: 19.23, 21.18 (CH_3), 39.64 (C^1H_2, Bn), 118.82 (C^2Cl_2), 127.51 (C^4H, Bn), 127.87 (C^3H, C^5H, Bn), 129.85 (C^3H, C^6H_3), 130.09 (C^5H, C^6H_3), 128.35 (C^2H, C^6H, Bn), 130.27 (C^6H, C^6H_3), 135.15 (C^2, C^6H_3), 135.48 (C^4, C^6H_3), 133.44 (C^1, C^6H_3), 134.10 (C^1, Bn), 138.95 (C^1). Mass spectrum, m/z (I_{re}l, %): 290/292/294 M^+ (16/10/2), 255/257 $[M–Cl^-]^+$ (69/22), 219 (100), 203(23), 177/179 (45/17), 149/151 (51/18), 119 (86), 101 (32), 91 (59), 77 (36), 65 (12), 51 (19).

A mixture of 1-(1-benzyl-2,2-dichlorovinyl)-2-chlorobenzene (VIe) and 1-(1-benzyl-2,2-dichlorovinyl)-4-chlorobenzene (VIf). Colorless liquid, bp 165–167°C (2 mm Hg). *ortho*-VIe. ^{13}C NMR spectrum, δ_C, ppm: 41.76 (C^1H_2, Bn), 119.93 (C^2Cl_2), 126.79 (C^4H, Bn), 129.52 (C^4H, C_6H_4), 127.93, 128.44 (C^3H, C^5H, C^6H_4), 128.02 (C^3H, C^5H, Bn), 128.29 (C^2H, C^6H, Bn), 128.38 (C^6H, C^6H_4), 138.74 (C^2, C^6H_4), 132.51 (C^1, C^6H_4), 137.46 (C^1, Bn), 134.76 (C^1). Mass spectrum, m/z (I_{rel}, %): 296/298/300/302 (45/36/12/2) $[M]^{+\cdot}$, 261/263/265 (44/23/4) $[M - Cl^\cdot]^+$, 225/227 (100/35), 183/185/187 (57/31/13), 149/151 (43/15), 136/138 (28/9), 125/127 (53/17), 95 (59), 90 (33), 75 (15), 63 (21), 51 (17). *para*-VIf. ^{13}C NMR spectrum, δ_C, ppm: 39.61 (C^1H_2, Bn), 119.27 (C^2Cl_2), 126.79 (C^4H, Bn), 128.02 (C^3H, C^5H, Bn), 130.12 (C^3H, C^5H, C^6H_4), 128.29 (C^2H, C^6H, Bn), 128.65 (C^2H, C^6H, C^6H_4), 138.42 (C^4, C^6H_4), 134.28 (C^1, C^6H_4), 135.63 (C^1, Bn), 138.24 (C^1). Mass spectrum, m/z (I_{rel}, %): 296/298/300/302 (45/36/12/2) $[M]^{+\cdot}$, 261/263/265 (44/23/4) $[M - Cl^\cdot]^+$, 225/227 (100/35), 183/185/187 (57/31/13), 149/151 (43/15), 136/138 (28/9), 125/127 (53/17), 95 (59), 90 (33), 75 (15), 63 (21), 51 (17).

KEYWORDS

- alkylation products
- compounds
- cyclopropanes
- hydrocarbons
- methyl substituents
- polyhalogen

REFERENCES

1. R. R. Kostikov, A. P. Molchanov, A. Ya. Bespalov. J. Org. Chem. (in Rus.), 1974, 10, 10.
2. N. S. Zefirov, I. V. Kazimirchik, K. L. Lukin. Science (in Rus.) 1985.
3. T. V. Arbuzova, A. R. Khamidullina, S. S. Zlotsky. Proceedings of higher educational establishments, Chemistry. Chem. Technol. (in Rus.), 2007, 50 (6), 15.
4. A. R. Khamidullina, E. A. Brusentsova, S. S. Zlotsky. Proceedings of higher educational establishments, Chemistry. Chem. Technol. (in Rus.), 2008, 51 (9), 106.

5. E. A. Kletter, Yu. P. Kozyreva, D. I. Kutukov, S. S. Zlotsky. Petrochemistry (in Rus.), 2010, 50, 65.
6. E. A. Brusentsova, S. V. Kolesov, A. I. Vorob'eva, et al. J. Org. Chem. (in Rus.), 2008, 78, 783.
7. A. R. Shiriazdanova, A. N. Kazakova, S. S. Zlotsky. J. Org. Chem. (in Rus.), 2009, 16, 142.
8. N. V. Zyk, O. B. Bondarenko, A. Yu. Gavrilova, et al. Proceedings of the Academy of Sciences, Chemistry (in Rus.), 2011, No. 2, 321.
9. A. N. Kazakova, L. V. Spirikhin, S. S. Zlotsky. Petrochemistry (in Rus.), 2012, V. 52, no. 2, 142.
10. N. V. Zyk., O. B. Bondarenko, A. Yu. Gavrilova, A. O. Chizhov, N. S. Zefirov, Proceedings of the Academy of Sciences, Chemistry (in Rus.), 2011, no. 2, 321.
11. H. Gunther, Introduction to the Course of NMR Spectroscopy. Springer-Verlag, 1984.

CHAPTER 15

GNRH ANALOGS AS MODULATOR OF LH AND FSH: EXPLORING CLINICAL IMPORTANCE

DEBARSHI KAR MAHAPATRA,[1] VIVEK ASATI,[2] and
SANJAY KUMAR BHARTI[2]

[1]School of Pharmaceutical Sciences, Guru Ghasidas
Vishwavidyalaya (A Central University), Bilaspur–495009, India;
Tel.: +91 7552–260027; Fax: +91 7752–260154;
E-mail: mahapatradebarshi@gmail.com

[2]School of Pharmaceutical Sciences, Guru Ghasidas
Vishwavidyalaya (A Central University), Bilaspur–495009,
Chhattisgarh, India

CONTENTS

15.1   INTRODUCTION

Gonadotrophin-releasing hormone (GnRH), a decapeptide composed of 10 amino acids, is secreted from the hypothal hypothalamus in a pulsatile manner. Through G protein-coupled receptors located on gonadotrope cells, GnRH stimulates the synthesis and release of the gonadotrophins FSH and LH, which in turn regulate folliculogenesis and steroidogenesis in the ovaries [1] (see Fig. 15.1). Furthermore, evidence suggests that GnRH may have a direct effect on the functions of the ovary, such as steroidogenesis, oocyte maturation and follicle rupture [2–5]. Many derivatives, known as GnRH analogs, including agonists and antagonists, were synthesized with an enhanced biological potency of binding with GnRH receptors [1]. These analogs have been used in ovarian stimulation to prevent a premature LH surge that can impair follicular development [6–7]. A GnRH agonist combined with FSH and/or human menopausal gonadotrophin (HMG), widely used in IVF centers, has some advantages in reducing the rate of cycle cancellation, improving the rate of live birth and enabling flexibility to schedule oocyte retrieval [8–10].

With the introduction of the gonadotrophin-releasing hormone (GnRH) antagonist protocol and the subsequent possibility of using a bolus of GnRH agonist (GnRHa) for final oocyte maturation and ovarian hyperstimulation syndrome (OHSS) prevention, the authors of this chapter founded a special interest group in 2009 'The Copenhagen GnRH Agonist Triggering Workshop Group.' The first meeting was held in Copenhagen,

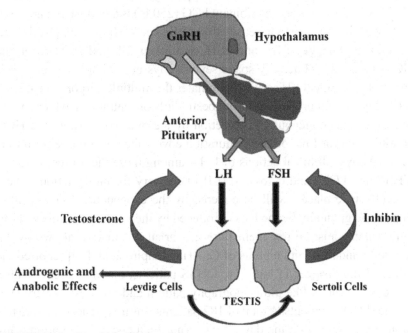

FIGURE 15.1 Steroid feedback pathway.

resulting in an extensive review [11], followed by several publications employing the GnRHa trigger concept. The purpose of most of the subsequent trials was to overcome the luteal-phase insufficiency previously reported post GnRHa trigger – despite supplementation with a standard luteal-phase support. A recent review focused on luteal-phase deficiency after ovarian stimulation in general, emphasizing the luteal phase after GnRHa trigger [12]. Following this publication, several studies were published on the subject. As the interest in GnRHa trigger is rapidly increasing, this review presents an update on the most recent literature and the continuing controversies regarding GnRHa trigger.

15.2 PHYSIOLOGY

LH homology, an extended half life and an easy manufacturing process turned human chorionic gonadotrophin (hCG) into an excellent molecule to be used for triggering of final oocyte maturation and ovulation during assisted reproduction treatment. Typically, one bolus of urinary HCG (5000–10,000 units) or recombinant hCG (250 lg) is administered approximately 36 h before oocyte retrieval in IVF cycles. Unlike the physiological mid-cycle surge of LH and FSH, terminating 48 h after its onset, the HCG-mediated LH activity spans several days into the luteal phase. This supra-physiological LH activity stimulates the multiple corpora lutea, leading to high serum progesterone and oestradiol concentrations, which in turn decrease the endogenous LH secretion by the pituitary [13]. As the HCG administered for final oocyte maturation covers the luteal phase for a total of 8–10 days, all luteal actions of LH – among those the up-regulation of VEGF and FGF2 and cytokines (LIF) necessary for implantation to successfully take place – will be covered by the exogenous hCG, and after this period gradually by the HCG produced by the implanting embryo [12]. GnRH antagonist co-treatment during ovarian stimulation allows ovulation to be induced with a bolus of GnRHa, Shapiro et al. [40] reported the effect of dual trigger in high-risk OHSS patients, mainly patients with a polycystic ovaries-like ultrasonography pattern and polycystic ovary syndrome (PCOS) patients. A total of 182 patients was treated according to this protocol, receiving on the day of triggering final oocyte maturation 4 mg leuprolide acetate as well as a mean of 1428 IU HCG for trigger. Patients had a mean of 28 follicles on the day of trigger and a mean of 20 oocytes were retrieved. After blastocyst transfer, an ongoing pregnancy rate of 58% per transfer was obtained. One patient (1/182) in this high-risk group of patients developed late-onset OHSS. Modified luteal-phase support post GnRH agonist trigger with HCG The basic premise behind this approach is to dissociate the ovulation trigger from the luteal support. While a bolus of GnRHa is responsible for an endogenous surge of LH and FSH, a low dose(s) of HCG after oocyte aspiration will replace the actions of early luteal LH to sustain implantation and luteal ovarian steroidogenesis.

This approach was previously reported prospectively by Humaidan et al. [15, 17] and Humaidan [16] in normo-responding as well as hyper-responding patients, proving that low-dose luteal HCG normalizes the reproductive outcome post GnRHa trigger. Moreover, similar results were recently reported by Radesic and Tremellen [18] who analyzed 71 consecutive high-risk OHSS cases, treated according to the protocol suggested by Humaidan et al. [17].

Forty-5 percent of patients were PCOS patients and the mean anti-Mullerian hormone concentration was 48 pmol/L. A high risk of developing OHSS was defined as the presence of at least 14 follicles 12 mm on the day of triggering final oocyte maturation. All patients were triggered with a bolus of GnRHa (leuprolide acetate, 2 mg), followed by 1500 IU HCG administered subcutaneously, after the oocyte retrieval. A mean of 17 oocytes were retrieved and all 71 patients received a single embryo transfer, which resulted in a clinical pregnancy rate of 52% per transfer and a miscarriage rate of 8%. One patient developed late-onset severe OHSS, requiring 7 days of hospitalization. Interestingly, a total of two injections of 1500 IU HCG – on the day of oocyte retrieval and 4 days later – in low-risk OHSS patients (less than 12 follicles >12 mm) rescued the luteal phase and the reproductive outcome to a degree that no additional luteal-phase support was needed [19].

This novel, simple and patient-friendly approach assures robust luteal activity as well as avoids lengthy luteal-phase treatment in the low-risk OHSS patient. as GnRHa displaces the GnRH antagonist in the pituitary, activating the GnRH receptor, resulting in a surge of gonadotrophins (flare-up), similar to that of the natural midcycle surge of gonadotrophins. However, significant differences exist between the GnRHa-induced surge and that of the natural cycle. Thus, the LH surge of the natural cycle is characterized by three phases, with a total duration of 48 h [20], as compared with the GnRHa-induced surge of gonadotrophins, which consists of two phases only, with duration of 24–36 h [21]. This leads to a significantly reduced total amount of gonadotrophins being released from the pituitary when GnRHa is used to trigger final oocyte maturation [21, 22].

Thus, the combined effect of ovarian stimulation and GnRHa trigger dramatically reduces the endogenous LH concentration during the early luteal phase [16], necessitating a modification of the standard luteal-phase support to secure the reproductive outcome [17]. At present, there is no commercial product that fully imitates the mid-cycle surge of gonadotrophins. An effort to introduce recombinant LH as a trigger was published in 1998 by the European Recombinant LH Study Group; however, it did not result in a commercial preparation due to the high dose of recombinant LH needed for trigger.

15.3   GnRH AGONIST

GnRH agonists are synthetically modeled ligands produced by modification in natural GnRH that are clinically useful for the treatment of various forms of cancer, delaying puberty in individual with precocious puberty and in delaying puberty pending in children with gender in congruency and in management of various forms of gynecological problems such as menorrhagia, ovarian hyperstimulation, uterine fibroids, endometriosis, etc. Uses are not limited to adults only rather transsexuals use them non-clinically for suppression of hormones.

15.4   CLASSIFICATION

GnRH agonists are synthetically prepared by specific substitution of amino acid, alkylation or deletion at position 6, 9 and 10 of natural GnRH structure. A list of approved GnRH agonists is mentioned in Table 15.1.

- 6-substituted analogs: Nafarelin, Tripotorelin;
- Analogues with single substitution: Histrelin, Deslorelin;
- Analogues with two substitution: Goserelin, Leuprolide, Buserelin.

TABLE 15.1   List of Approved GnRH Agonist With Clinical Uses

GnRH Agonist	Clinical uses
Leuprolide	• For management of endometriosis, where main foci is pain management and reduwction of lesions. Norethindrone acetate is used in combination for management of recurrence of symptoms. • Treatment of advanced prostatic cancer.
Goserelin	• For management of endometriosis, where main foci is pain management and reduction of lesions. Norethindrone acetate is used in combination for management of recurrence of symptoms. • Treatment of advanced prostatic cancer in combination with antiandrogens. • Treatment of advanced breast cancer. • As endometrial-thinning agent for management of dysfunctional uterine bleeding.
Histrelin	Treatment of advanced prostatic cancer
Triptorelin	Treatment of advanced prostatic cancer
Nafarelin	Treatment of advanced prostatic cancer
Deslorelin	Treatment of advanced prostatic cancer
Buserelin	Treatment of advanced prostatic cancer

15.5   LEUPROLIDE

Leuprolide as acetate salt is a synthetic GnRH agonist clinically approved for diverse treatment of prostate cancer, central precocious puberty, endometriosis, uterine fibroids, in vitro fertilization, polycystic ovary syndrome, functional bowel disease, contraception, premenstrual syndrome and Alzheimer disease [23]. The basic mechanism of action in each case involves suppressing gonadotrope secretion of LH and FSH that suppresses gonadal sex steroid production by interacting with the receptors called GnRHR in cortex and hippocampus areas of the brain. With the suppression of serum gonadotropes, all clinical manifestations get suppressed. After initiating therapy with leuprolide, a marked increase in serum hormonal level of LH, FSH, sex steroids is observe, that later falls with time of 2–4 weeks due to "down regulation" of receptors. This can be compare similar to castration level in male and postmenopausal stage in female [24].

Leuprolide acetate is a well-tolerated drug but when injected in body, it produces a sudden hike in hormonal level, which leads to appearance of secondary sexual characteristics in boys and initiation of menses in girls. In case of endometriosis and uterine fibroid, heavy bleeding and tumor growth is observed. Although some general side effects are often observed such as nausea, amenorrhea, bone pain, decreased libido, insomnia, headache, impotence and hot flush [25].

15.6 TRIPTORELIN

Triptorelin is a synthetic agonist characterized by substitution of glycine in 6th position by D-tryptophan, which increases its affinity and provides resistance to enzymatic degradation. It is available in two formulations: short acting daily shot of 0.1 mg of agonist and a long acting 4 week formulation of 3.75 mg. Both are used as a hormonal therapy in the treatment of advanced prostate cancer by intramuscular or subcutaneous route [26]. Triptorelin find its another great importance in preventing chemotherapy induced early menopause in women with breast cancer. This hormonal therapy comes in depot form for reduction of LHRH level under trade name Gonapeptyl® and sustained release formulation is available under trade name of Decapeptyl®. As mentioned earlier, all GnRH agonists show a characteristic rise in serum hormonal levels, which "flares up," and precipitate tumor growth, initiates menses, etc. With administration of Triptorelin, an initial surge of Testosterone is observed; the testosterone is biochemically converted to its more active form known as Dehydrotestosterone in presence of enzyme 5-α-reductase which promotes faster tumor growth [27]. This is blocked by using a combination of Triptorelin with Antiandrogens (such as Bicalutamide, Flutamide, etc.), which blocks the enzyme to prevent cancerous growth in prostate.

Triptorelin

15.7 TRIPTORELIN

Triptorelin causes varieties of side effects depending upon the dose administered. Some patient experiences fewer side effects whereas others reported to have a large number of side effects and associated interactions. These drugs are administered intentionally to reduce the level of hormones, but these hormones have anabolic effects which enable formation of all micro and macro physiological components in body such as bone calcification, bone formation, mood elevation, sleep physiology, etc. With the disruption of normal hormonal function, some common side effects occur such as loss of libido, osteoporosis, depression, tiredness, gynaecomastia, sore joints, etc. [28].

15.8 HISTRELIN

Histrelin in acetate salt is a 200 times potent agonist preferred exclusively for the treatment of central precocious puberty, a one of the rare disease caused by an activation of hypothalamic-pituitary-gonadal axis, which results in an early manifestation of puberty in children at age of 7–8 years in girls and 9 years in boys. The disease is characterized by elevated LH and FSH in response to GnRH stimulation, which influences rapid progression of pubertal development and skeletal maturation [29]. Histrelin, when administered by nasal, subcutaneous or intramuscular route acts by down regulating GnRH receptors, which inhibits gonadotropin release and therefore this, leads to suppression of hypothalamic-pituitary-gonadal axis. The clinical result is slowing of bone age advancement as well as growth velocity and pubertal manifestation. The clinical effects produced by GnRH agonists are in order: Leuprolide < Buserelin < Deslorelin < Nafarelin < Histrelin. Histrelin acetate is available as flexible, non-biodegradable, biocompatible hydrogel implants with microporous wall, which releases the drug continuously at a rate of 50–65 µg/day. Implants are first approved by USFDA for the treatment of central precocious puberty under the trade name Supprelin®. The drug core contains inactive diluent stearic acid. The hydrogel reservoir is a hydrophilic polymer cartridge composed of 2-hydroxyethyl methacrylate, 2-hydroxypropyl methacrylate, trimethylolpropane trimethacrylate, benzoin methyl ether, Perkadox-16, and Triton X-100. The hydrated implant is packaged in a glass vial containing 2.0 mL of 1.8% NaCl solution [30].

15.9 HISTRELIN

In human, administration of histrelin leads to an initial increase in LH and FSH, which leads to increase in Testesterone. But, continuous administration of histrelin for a month results in decreased levels of FSH and LH. Histrelin is classified as category X drug for pregnancy condition. Animal models show major fetal abnormalities in rabbit throughout the gestation. The animal showed fetal mortality, decreased fetal weights and is expected to show prompt abortion. On continuing the pharmacotherapy with this agonist particularly shows fatal effects such as urinary obstruction, hyperglycemia and increased

Histrelin

risk of diabetes mellitus, hematuria, myocardial infarction, stroke, etc. Less severe complications include tenderness, swelling, erythema, whirl and flare [31].

15.10 DESLORELIN

Deslorelin is an agonist popular in mediating contraception. It is marketed under trade name Suprelorin as implants for male dogs to prevent fertility for a temporary period. When this implant is inserted subcutaneously under skin it works effectively for 6 months at dose of 4.7 mg and for 12 months at dose of 9.4 mg in suppressing fertility, when gets activated after 6 weeks. Deslorelin acts like the natural hormone GnRH, which controls the secretion of other hormones, involved in fertility. Suprelorin is given as an implant that slowly releases a continuous low dose of deslorelin, which suppresses the production of FSH and LH. As a result in male dogs less testosterone circulates in the blood, the dog stop producing sperm and its libido is reduced [32]. Committee for Medicinal Products for Veterinary Use (CVMP) approves this drug for use in male dog and concludes several potential advantages, which outweigh risks for inducing infertility. On initiating the treatment, following major desired features are observed: reduction in spermatogenesis, reduction in testicle size and reduction in libido. When treatment stops, the condition reverses and successful mating is reported. Some of the associated risks include moderate swelling, local inflammation, etc. [33, 34].

Deslorelin

15.11 GOSERELIN

Goserelin is a drug used exclusively for treatment of breast cancer. It works by suppressing the level of estrogen, which helps in growth of cancer cells. After injecting for 20 days, an acute fall in level of estrogen is observed which shows similar symptoms compared to postmenopausal women by firstly, acting on GnRH receptor to block gonadotrope mediated rise in LH & FSH which ultimately triggers the level of estrogen in body. But, once discontinued the level of estrogen rises again and thus it is continued for life [35]. Goserelin comes in implant in a small syringe that delivers the drug via subcutaneous route in abdomen, which provides a "depot" release. For primary cancer, it is repeated after each month [36].

Goserelin

15.12 GOSERELIN

Other clinical applications of goserelin includes primary drug for endometriosis and uterine fibroid, where after a single shot it acts by reducing estrogenic levels and stops menses completely compared to Leuprolide where an initial "flare" of estrogen is observed which triggers heavy menses. Even after discontinuing the drug, no menses are observed even after 6–12 months. Goserelin is considered as safe drug, still it show side effects such as hot flushes, mood changes and decrease in libido [37].

15.13 NAFARELIN

Nafarelin is used for the treatment of treatment of estrogen-dependent conditions such as endometriosis or uterine fibroid, in treatment of central precocious puberty, menstrual cramp, painful intercourse, pelvic pain and in controlling ovarian stimulation in IVF. It is delivered twice daily as nasal formulation under trade name Synarel®. It is highly effective in ceasing heavy menstrual flow by blocking LH and FSH induced estrogen production. Nafarelin shows a range of side effects such as nasal irritation, hot flushes, gynaecomastia, mood swings, headache, skin irritation, insomnia, muscle ache, rhinitis, etc. [38].

Nafarelin

15.14   BUSERELIN

It is similar to Nafarelin in all aspects: mechanism of action, clinical use and mode of administration. Although subcutaneous injection is also available but nasal formulation Suprefact® is more popular. It finds its role in ovulation induction by inducing pituitary blockade [39].

Buserelin

15.15 DUAL TRIGGER

'Dual trigger' is the concept in which the benefits of a bolus of a GnRHa in terms of release of endogenous LH and FSH from the pituitary are combined with the long acting LH activity of a small bolus of HCG, covering the early luteal-phase LH deficiency, previously described after GnRHa trigger. The dual trigger protocol is usually followed by a standard luteal-phase support.

Shapiro et al. [40] retrospectively reported the effect of dual trigger in high-risk OHSS patients, mainly patients with a polycystic ovaries-like ultrasonography pattern and polycystic ovary syndrome (PCOS) patients. A total of 182 patients was treated according to this protocol, receiving on the day of triggering final oocyte maturation 4 mg leuprolide acetate as well as a mean of 1428 IU HCG for trigger. Patients had a mean of 28 follicles on the day of trigger and a mean of 20 oocytes were retrieved. After blastocyst transfer, an ongoing pregnancy rate of 58% per transfer was obtained. One patient (1/182) in this high-risk group of patients developed late-onset OHSS. Modified luteal-phase support post GnRH agonist trigger with HCG.

15.16 MODIFIED LUTEAL-PHASE SUPPORT POST-GnRH AGONIST TRIGGER

Babayof et al. [41] were the first to report the use of intensive oestradiol and progesterone for luteal support in high-risk OHSS patients post GnRHa trigger. The reproductive outcome of this trial was disappointingly low. Engmann et al. [42] used a similar approach in high-risk OHSS patients (n = 33) post GnRHa trigger reporting the prevention of OHSS while maintaining good pregnancy rates. In contrast, the OHSS rate in the HCG trigger control group (n = 32) was 30%. These findings were recently supported by Imbar et al. [43] in an observational trial in high-risk OHSS patients comparing the outcome of fresh transfer (n = 70) after GnRHa trigger and an intensive luteal-phase support with oestradiol and progesterone, based on Engmann et al. (2008), versus frozen–thawed embryo transfer cycles (n = 40). The authors reported a live birth rate of 27% after fresh transfer. Orvieto [44] published a report including a cohort of 67 high-risk OHSS patients treated during the period 2010–2011, who after GnRHa trigger received an intensive luteal-phase support with oestradiol and progesterone, similar to the one reported by Engmann et al. [45]. However, despite the intensive luteal support, implantation and pregnancy rates remained disappointingly low-comparable to results previously published after GnRHa trigger without intensive luteal support [45]. This finding was commented by Benadiva and Engmann [46] in a Letter to the Editor with updated recommendations for handling of the luteal phase after GnRHa trigger. Griffin et al. [47] reported that patients should be stratified according to their oestradiol concentration on the day of triggering final oocyte maturation. In patients with oestradiol concentrations >4000 pg/mL, the authors adhere to the previously published protocol of intensive luteal-phase support, only. In patients with oestradiol concentrations <4000 pg/ml a dual trigger is used (GnRHa + 1000 IU HCG) in combination with intensive luteal-phase support.

15.17   OHSS

A recent report by ESHRE by de Mouzon et al. [48] clearly reveals that OHSS is still one of the major complications of ovarian stimulation for IVF. A diagnosis of OHSS was reported in 26 of the 33 countries. In total, 2470 cases of OHSS were recorded in 2007, corresponding to an OHSS occurrence of 0.7% in all stimulated cycles (0.8% in 2006). A wide variation in the occurrence of OHSS was reported: UK (1.3%) and Russia (1.8%) having the highest rates, and Germany and Spain the lowest (0.3%). However, the issue of OHSS reporting still seems to be a 'gray zone' with major under-reporting. Importantly, OHSS may have a lethal outcome. The UK-based 'Confidential Enquiry into Maternal and Child Health' organization has reported four OHSS-related deaths during the years 2003–2005, which translates into about three OHSS-related deaths per 100,000 assisted reproduction cycles in the UK. Data from the Netherlands by Braat et al. [49] reported a similar incidence. Importantly, all OHSS-related deaths in the latter report were seen after HCG trigger of final oocyte maturation, following the implementation of a 'freeze-all' strategy for OHSS prevention.

15.18   OHSS AND GnRHa TRIGGER

Virtually complete elimination of OHSS is one of the major benefits of GnRHa trigger. The mechanism behind this phenomenon is the luteolysis induced by a bolus of GnRHa caused by the short half-life of endogenous LH as compared with HCG. Once the endogenous HCG production from the trophoblast reaches measurable serum concentrations around day 8 after ovulation reported by Bonduelle et al. [50], it is too late to rescue the corpora luteae, which results in virtual elimination of the late-onset pregnancy-associated OHSS [51, 52]. Taken together, the combination of GnRH antagonist co-treatment and GnRHa trigger is the tool by which the concept of a future OHSS-free clinic could become a reality [53, 54]. So far, only one case of severe, early onset OHSS after GnRHa trigger was reported in the English literature [55]. This case, therefore, merits closer scrutiny. A 30-year-old patient with PCOS and male factor infertility underwent her first stimulated cycle. The oestradiol concentration on the GnRHa trigger day was 47,877 pmol/l (13,041 pg/ml) and 13 oocytes were retrieved. Following oocyte retrieval the patient was hospitalized with abdominal distension, enlarged ovaries and lower abdominal pain. She received low molecular weight heparin, cabergoline (0.5 mg/day) and i.v. infusion therapy, including albumin. Due to a drastic decrease of the hemoglobin concentration to 4.9 mmol/L (8 g/dl), the patient received blood transfusion 2 days post oocyte retrieval. Importantly, the haematocrit concentration was 0.41 on the GnRHa trigger day, 0.37 on the oocyte retrieval day and <0.35 post blood transfusion. Three to four days post trigger, 3.9 liters of 'blood-stained ascites' were drained, indicative of a subacute intraperitoneal hemorrhage – a well-known complication of vaginal ultrasound-guided oocyte retrieval; unfortunately, in the conclusions of the abstract, this was described as .'.a single case of a severe early onset OHSS' even though this diagnosis was clearly not supported by the clinical details since the hallmark of OHSS is haemo-concentration.

15.19   COCHRANE REVIEWS AND GnRHa TRIGGER

A recent Cochrane review comparing GnRHa versus HCG for triggering of final oocyte maturation in IVF concluded that GnRHa as a trigger of final oocyte maturation should not be routinely used due to the significantly lower live birth rate [56]. Moreover, the section on implications for research stated: 'In view of the poor reproductive outcomes following oocyte triggering with GnRH agonist we believe there is no indication for further research with GnRH agonists for oocyte triggering in ART in fresh autologous cycles.' This conclusion was unprecedented and obviously too premature, based on an analysis compiling data from some of the initial clinical trials, reporting a very poor reproductive outcome, with newer trials in which modifications of the luteal-phase support resulted in pregnancy rates comparable to those seen after HCG trigger – and with a reduction in the OHSS rate. Thus, apples were compared with oranges. This issue was thoroughly addressed in debates questioning not only the performance of meta-analyzes during the development of new concepts, but also the role of meta-analyzes as a whole in the field of reproductive medicine [54, 57].

15.19.1   PREDICTIVE FACTORS AND HOW TO HANDLE PATIENTS AT RISK OF OHSS

Predicting OHSS is difficult; there are, however, predictive factors, which need to be taken into consideration when choosing the type of GnRH analog, the FSH dose and the duration of stimulation as well as the choice of trigger method. Predictive factors for OHSS may be divided into primary and secondary risk factors. The most important primary risk factors are: high antiMullerian hormone, high antral follicle count, PCOS, isolated PCOS characteristics and a previous history of OHSS. In contrast, an important secondary risk factor is the number of follicles >11 mm on the day of trigger, and a threshold of more than 14 follicles >11 mm has previously been shown to predict 87% of severe OHSS cases [58]. Once a decision to use GnRHa trigger has been made, there are now several clinical options.

15.20   FRESH TRANSFER AND LOW-DOSE HCG SUPPLEMENTATION TO SECURE THE REPRODUCTIVE OUTCOME

This concept has until now resulted in a non-significant difference in delivery rates and a significant reduction in OHSS rate in the high-risk OHSS patient. The concept was recently used in a prospective randomized trial including a total of 118 high-risk OHSS patients (14–25 follicles) with no OHSS development in the GnRHa triggered group versus 3% in the HCG-triggered group. Above 25 follicles, we recommend either a 'freeze-all' policy or intensive luteal-phase support with oestradiol and progesterone. A fresh transfer in combination with a dual trigger (1000 IU HCG) and luteal support with oestradiol and progesterone in patients with oestradiol concentrations <4000 pg/ml on trigger day [19, 43, 47].

The use of GnRHa trigger will prevent OHSS even in extreme cases if a freeze-all policy is adopted [54]. With the current improvement in cryo-technology, an excellent OHSS risk-free cumulative pregnancy rate has previously been reported [55]. GnRHa trigger is the trigger method of choice for the oocyte donor due to several advantages over HCG trigger, among those: virtual elimination of OHSS, a reduced luteal ovarian volume leading to diminished abdominal distension and pain, and a short interval until withdrawal bleeding occurs; factors that substantially decrease the treatment burden of the oocyte donor [59]. Moreover, GnRHa trigger should be considered in situations like repeated IVF failure, empty follicle syndrome and repeated retrieval of immature oocytes as a subset of patients may require the FSH surge – in addition to the LH surge – to promote final oocyte maturation [60].

15.21 GONADOTROPHIN RELEASING HORMONE (GNRH) ANTAGONIST

Gonadotrophin Releasing Hormone Antagonist are a class of peptide molecules produced synthetically by iterative synthesis of multiple amino acid units. They are similar in structure to that of natural hormone GnRH, produced in neuronal section of hypothalamus. The first generation drugs had lower potency and high histamine releasing potential. The second generations have little increase in potency after incorporation of a D-amino acid at 6-position but anaphylactic reactions still prevailed leading to gel clot formation. The third generation molecules proved efficacious after replacement of D-Arg at 6-position by ureidoalkyl amono acids [61]. After administration ~70% suppression in Luteinizing Hormone (LH) and ~30% suppression in Follicle Stimulating Hormone (FSH) were observed [62].

GnRH Antagonist binds with gonadotrophin releasing hormone (GnRH) receptors in the pituitary and blocks the interaction gonadotrophin releasing hormone, thus blocking luteinizing hormone (LH) and Follicle Stimulating Hormone (FSH) release. The resulting consequence is an immediate scarcity of luteinizing hormone (LH), which leads to a periodic blockade of testosterone release from testes [63]. A list of some common GnRH antagonists is mentioned in Table 15.2.

15.22 CLASSIFICATION

- First Generation: Azaline B.
- Second Generation: Degarelix.
- Third Generation: Abarelix, Cetrorelix, Ganirelix, Iturelix, Antarelix.
- Fourth Generation: Ozarelix.

15.23 DEGARELIX

Degarelix is a synthetic peptide derivative of natural gonadotropin releasing hormone (GnRH) decapeptide which is available commercially in acetate salt form, is a prescription medication approved to treat men having advanced stages of prostate cancer [64]. It is a linear decapeptide amide containing seven artificial amino acids, five of which are D-amino acids. It was first manufactured by Rentschler Biotechnologie for Ferring

TABLE 15.2   List of Some Common GnRH Antagonists With Chemical Properties

	Degarelix	Ganirelix	Abarelix	Cetrorelix
Bioavailability (%)	30–40	90–91	90–92	80–85
Protein Binding (%)	90	82	96	86
Molecular Formula	$C_{84}H_{107}ClN_{18}O_{18}$	$C_{80}H_{113}ClN_{18}O_{13}$	$C_{72}H_{95}ClN_{14}O_{14}$	$C_{70}H_{92}ClN17O_{14}$
Molecular Mass (g/mol)	1692.31	1570.4	1416.06	1431.06

pharmaceuticals Limited. The Food and Drug Administration (FDA) of United States of America (USA) approved on 24th December, 2008 for use. On 17th February, 2009 it was further approved by the European Medicines Agency (EMEA) for use in adult male only. Earlier castration was the only way to reduce testosterone level circulating in the body as a part of tumor growth management in advanced prostate cancer stages, as it enables tumor proliferation and growth. Degarelix proved to be a better option for tumor management with reduced side effects and absence of Hypothalamic-Pituitary-Gondal-Axis (HGPA) stimulation makes it clinically safer to use. The product comes under two formulations Subcutaneous Injection of 80 mg and 120 mg per vial and is under distribution by Australia, Canada, Czech Republic, Israel, Japan, New Zealand, Republic of China, Republic of Korea, Singapore, Slovakia, Turkey and United States of America. In healthy men, approximately 20–30% of an administered IV dose of degarelix was excreted unchanged in the urine [65].

The *in vitro* metabolite 70–80% of an administered dose is excreted as degarelix cleavage products via the hepato-biliary system. The majority of the remainder of the administered dose appeared to be eliminated by the hepato-biliary system as degarelix cleavage products. *In vitro* metabolism of degarelix at a concentration of 40.4 μM by CYP450 isoenzymes was investigated in human liver microsomes. The test concentration was approximately three orders of magnitude higher than the plasma concentration of degarelix obtained 12 hours after a SC injection of 40 mg in healthy males. Six metabolites were detected and five of these were oxidative metabolites of degarelix. The total amount of oxidative metabolites detected was low (<1% of the initial amount of degarelix) indicating that degarelix is likely to be a poor substrate for human CYP450 isoenzymes. The sixth metabolite (~2% of the initial amount of degarelix) was probably formed by proteases and not by CYP450 isoenzymes. Firmagon product is the intrinsic tendency of degarelix to gel at concentrations above about 1 mg/mL. The proposed products are administered as 40 mg/mL or 20 mg/mL solutions. While aqueous solubility of degarelix in water is initially high, the solutions turn turbid or viscous on standing. Thus Ferring describes the reconstituted product as a suspension. This is attributed to fibrillation (self-aggregation) of the peptide. *In vivo* gelling gives a depot formation instantaneously following subcutaneous administration. Firmagon is a powder for injection; it is administered by abdominal, subcutaneous injection after reconstitution with Water for Injection [66, 67].

A starting dose of 240 mg (administered as 2 × 120 mg) is recommended, followed by a monthly dose of 80 mg. Ferring [17] proposes registration of both 80 and 120 mg powders for injection. The products are presented with diluent (water for injection in a vial), syringes and needles. The starting and monthly doses are administered as different concentration solutions. The presentation of the diluent is perhaps suboptimal. With the 'starting dose pack,' 3.0 mL is taken from each of the two 6 mL diluent vials (remainders discarded). With the 'maintenance dose pack,' 4.2 mL is taken from one 6 mL diluent vial. Studies supported that the starting dose is 240 mg administered as two s.c. injections of 120 mg at a concentration of 40 mg/mL along with the maintenance dose of 80 mg administered as one s.c. injection at a concentration of 20 mg/mL [64, 68].

Side effects and ADR are less & limited but are very significant. It is found to prolong QT/QTc interval as long-term androgen deprivation therapy may prolong the QT interval, which is evidenced. Hypersensitivity reactions include anaphylaxis, angioedema, or severe cutaneous skin reactions related to degarelix treatment have been observed. A long-term usage show changes in bone density, decreased bone density has been reported in the medical literature in men who have had orchiectomy or who have been treated with a GnRH agonist. Anti-degarelix antibody development has been observed

Degarelix

in 10% of patients after treatment. Patients with known or suspected hepatic disorder have not been included in long-term clinical trials with degarelix. Mild, transient increases in ALT and AST have been seen [67, 69].

15.24 SYNTHESIS

The synthesis of degarelix acetate employed iterative peptide coupling and protection/de-protection sequences to obtain a yield of 85–99%, Boc-D-alanine [1] was immobilized via MBHA resin by reaction with diiso-propyl carbodiimide (DIC) and 1-hydroxybenzotriazole (HOBT). The resulting product was treated with trifluoroacetic acid (TFA) to remove the N-Boc protecting group to reveal amine [2]. The N-terminus of [2] was then subjected to sequential coupling and de-protection cycles with the following protected amino acids: N-Boc-L-proline, N-a-Boc-N^6-isopropyl-N^6-carbobenzoxy-L-lysine and N-Boc-L-leucine to give [3] and [4], respectively. The N-terminus of [4] was coupled with N-a-Boc-D-4-(Fmoc-amino)phenylalanine, followed by removal of the Fmoc group with piperidine in DMF to give the corresponding free aniline. The free aniline resin was then reacted with t-butyl isocyanate to generate the corresponding t-butyl urea followed by reaction with TFA to remove the Boc group to give the t-butyl urea amine [5]. The N-terminus of [5] was coupled with N-a-Boc-L-4-(Fmoc-amino)phenylalanine, followed by removal of the Fmoc group with piperidine in DMF to generate the corresponding free aniline. The free aniline was reacted with L-hydroorotic acid, followed by reaction with TFA to liberate amine [6]. Amine [6] was then coupled with O-benzylated-N-Boc-serine, followed by removal of the Boc group with TFA and reacting the resulting amine with N-a-Boc-D-(3-pyridyl)ala-nine and subsequent removal of the Boc group with TFA gave amine [7]. Amine [7] was coupled with N-Boc-D-(4-chlorophenyl) alanine, followed by removal of the Boc group with TFA, and the resulting amine was then coupled with N-Boc-D-(2-naphthyl) alanine, followed by removal of its Boc group with TFA to give [8]. Acylation of [8] with acetic anhydride fol-lowed by sequential treatment with HF and TFA resulted in cleavage from the resin, removal of the O-benzyl group, and conversion of the t-butyl urea to the corresponding NH2-urea, resulting in free degarelix. Finally, treatment with acetic acid provided degarelix acetate [9, 70].

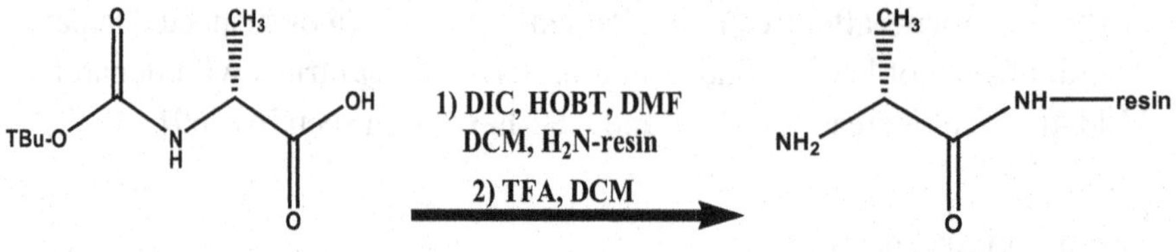

[1]

1) DIC, HOBT, DMF
   DCM, H₂N-resin
2) TFA, DCM

[2]

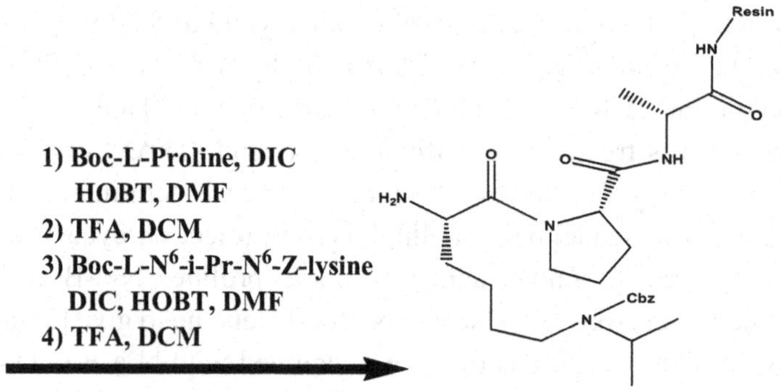

1) Boc-L-Proline, DIC
   HOBT, DMF
2) TFA, DCM
3) Boc-L-N⁶-i-Pr-N⁶-Z-lysine
   DIC, HOBT, DMF
4) TFA, DCM

[3]

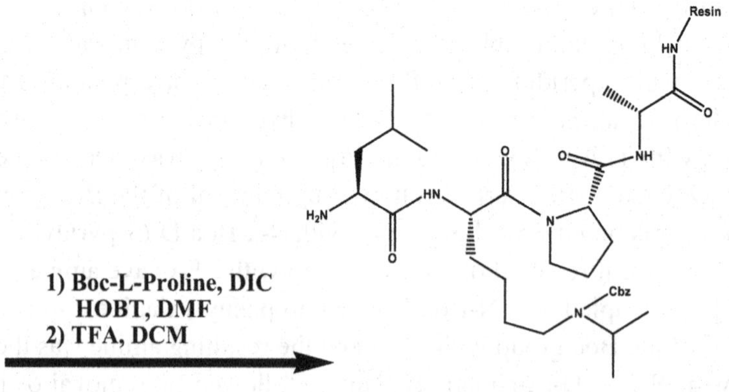

1) Boc-L-Proline, DIC
   HOBT, DMF
2) TFA, DCM

[4]

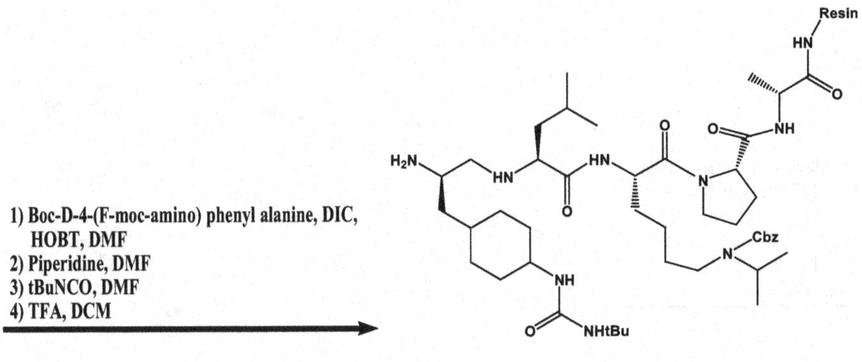

1) Boc-D-4-(F-moc-amino) phenyl alanine, DIC,
   HOBT, DMF
2) Piperidine, DMF
3) tBuNCO, DMF
4) TFA, DCM

[5]

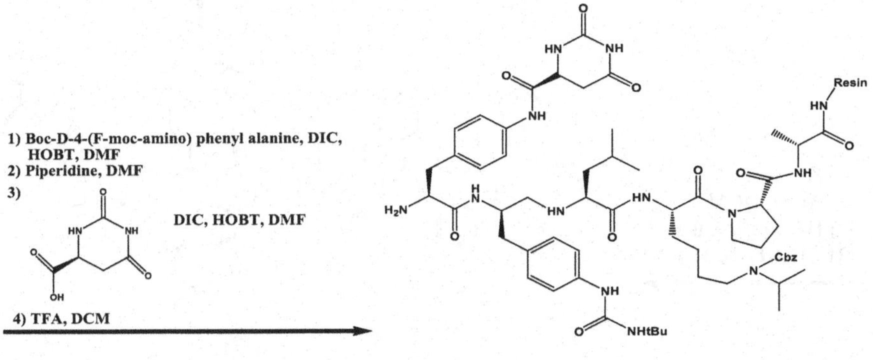

1) Boc-D-4-(F-moc-amino) phenyl alanine, DIC,
   HOBT, DMF
2) Piperidine, DMF
3)

DIC, HOBT, DMF

4) TFA, DCM

[6]

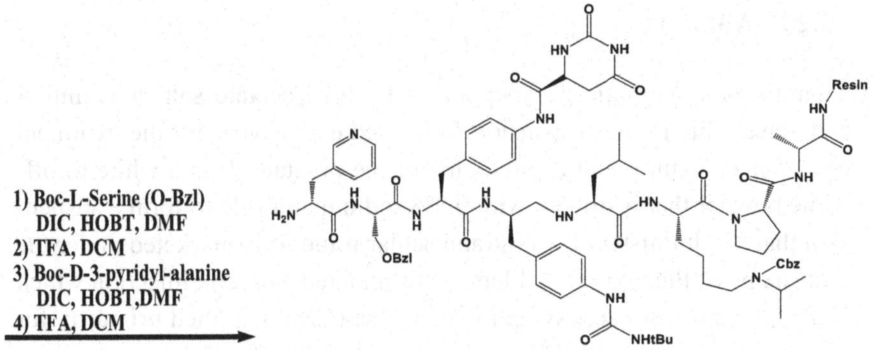

1) Boc-L-Serine (O-Bzl)
   DIC, HOBT, DMF
2) TFA, DCM
3) Boc-D-3-pyridyl-alanine
   DIC, HOBT,DMF
4) TFA, DCM

[7]

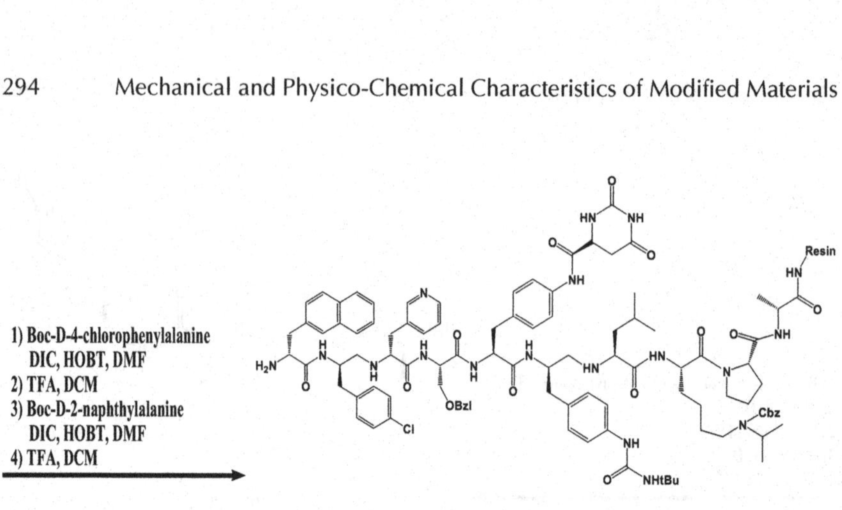

1) Boc-D-4-chlorophenylalanine
DIC, HOBT, DMF
2) TFA, DCM
3) Boc-D-2-naphthylalanine
DIC, HOBT, DMF
4) TFA, DCM
Resin
Cbz
OBzl
NHtBu

[8]

1) Ac₂O, DCM
2) HF, anisole, 0 c
3) CH₃CO₂H, H₂O
. CH₃CO₂H

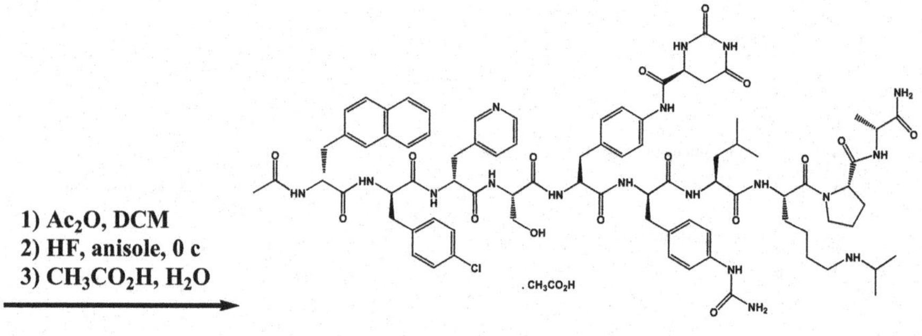

[9]

15.25 ABARELIX

Abarelix is a synthetic decapeptide in form of acetate salt, is commercially available in the suspension form used exclusively for the treatment of advanced symptomatic carcinoma of the prostate. It is a white to off-white powder that is diluted with 0.9% sodium chloride to form a suspension that is administered via intramuscular route. It is marketed under the trade name of Plenaxis [71]. Plenaxis formulated Abarelix injection which is complexed with carboxymethyl cellulose (CMC) applied primarily for long duration with depot release characteristics. The formulation is administered on Day 1st, 15th, 29th and every 4th weeks [72, 73].

Abarelix

15.26   CETRORELIX

Cetrorelix is a synthetic decapeptide analog of naturally occurring gonad-orelin, is a gonadorelin antagonist. It immediately suppresses gonadotro-pins and testosterone secretion without stimulatory effect. It competitively blocks gonadorelin receptors on the anterior pituitary gonadotroph and the subsequent transduction pathway, inducing a rapid, reversible sup-pression of gonadotropin secretion administered *SC* 250 mcg/day, given either in the morning beginning on the day 5 or 6 of ovarian stimulation or in the evening beginning on day 5, and continued until ovulation induc-tion [74]. As acetate salt it is equivalent to 0.25 mg cetrorelix either in the morning beginning on the day 5 or 6 of ovarian stimulation or in the evening beginning on day 5, and continued until ovulation induction. A single dose equivalent to 3 mg cetrorelix on day 7; if follicle growth does not allow ovulation induction within 4 days, additional doses of 250 mcg once daily may be given until the day of ovulation induction. Some com-mon associated ADR's are hypersensitivity reaction, moderate to severe renal or hepatic impairment [75]. Women are prone with severe allergic conditions during pregnancy and lactation. Children (below 14 years) and elderly people (above 65 years) are reported to have allergic symptoms. Other ADR's include mild and transient reactions at injection site, nausea, headache, ovarian hyperstimulation syndrome, systemic hypersensitivity reactions [76]. Women with active allergic conditions or a history of aller-gies needs precaution during treatment regimen. Cetrorelix is a Category X drug (as scheduled by USFDA). Studies in animals or human beings have demonstrated fetal abnormalities or there is evidence of fetal risk based on human experience or both, and it is demonstrated that the risk of using the drug in pregnant women clearly outweighs any possible benefit. Therefore the drug is contraindicated in women who are or may become pregnant [77].

Cetrorelix

15.27   OZARELIX

Ozarelix Acetate is a fourth generation decapeptide used primarily for the treatment of Hormone-dependent prostate cancer administered by intramuscular or subcutaneous route act by reversibly and dose-dependently suppresses gonadotropin and sex steroid levels by inhibiting the binding of endogenous LHRH to its receptors [78]. Ozarelix has the potential to be an important addition in treating hormone-dependent prostate cancer patients due to its ability to induce prolonged testosterone suppression without inducing an initial testosterone surge and associated risk of clinical flare and cancer progression observed in currently marketed LHRH agonists, as shown in healthy volunteers in early trials and are used along with Degarelix [79]. Spectrum Pharmaceuticals had enrolled 150 patients into a Phase IIb trial of a new LHRH receptor antagonist called ozarelix (a drug with a similar mechanism of action to degarelix or Firmagon) for the treatment of advanced forms of prostate cancer. The trial was performed to assess the safety and efficacy of monthly dosing with ozarelix (administered subcutaneously) compared to treatment with a depot for-mulation of goserelin acetate. Two patients will be randomized to receive ozarelix for each patient randomized to receive goserelin acetate. Patients will be followed for a total of 84 days, and the primary outcome will be the percentage of patients who achieve a serum testosterone level ≤ 50 mg/dl from day 28 to 84 [80].

Oxarelix offers certain advantages compared to other drugs of this class; they are, dose-dependent suppression of testosterone levels, improved solu-bility compared to other antagonists in the same class and no risk of tes-tosterone surge or clinical flare. Apart from its successful application in treatment of prostate cancer, it is also used for treatment of Benign Prostatic Hyperplasia (BPH), Endometriosis and Ovarian cancer. This product is offered and sold solely for uses reasonably related to the development and submission of information under a Federal law which regulates the manu-facture, use or sale of drugs (the "Bolar Exemption"). Bachem cannot be made liable for any infringement of intellectual property rights. It is the sole and only responsibility of the purchaser or user of this product to com-ply with the relevant national rules and regulations [81].

Ozarelix

15.28 GANIRELIX

Ganirelix acetate or diacetate is an injectable drug administered by subcu-taneous injection of 250 μg once per day during the mid to late follicular phase of a woman's menstrual cycle, is marketed universally by Organon

Internationals under the trade name Antagon. Ganirelix is derived from GnRH itself, with amino acid substitutions made at positions 1, 2, 3, 6, 8, and 10 [82]. Ganirelix is primarily used in assisted reproduction to control ovulation rather for prostate cancer suppression. The drug works by blocking the action of GnRH upon the pituitary, thus rapidly suppressing the production and action of LH and FSH. They are used in fertility treatment to prevent premature ovulation that will result in harvesting of eggs that are too immature for application in in vitro fertilization. Nearly all antagonists can be used for this purpose, Ganirelix is often primarily used, but the clinical success is the sole factor for primary preference of a drug. Thus prevents premature ovulation in women undergoing fertility treatment involving ovarian hyperstimulation that may cause the ovaries to produce multiple eggs. When such premature ovulation occurs, the eggs released by the ovaries may be too immature to be used in in-vitro fertilization. Ganirelix prevents ovulation until it is triggered by injecting human chorionic gonadotrophin [83]. Some of the common side effects include redness and swelling at the site of injection, abdominal pain and vaginal bleeding [84].

Ganirelix

15.29 ANTARELIX

Antarelix or EP 24332 is a GnRH antagonist that is reported to highly water soluble and highly potent with least histaminic discharge which makes it clinically useful for application in in vitro fertilization [85]. Antarelix is effective in influencing the GnRH-stimulated LH secretion of pituitary cells in vitro. After administration of Antarelix in vivo, the GnRH-stimulated LH secretion of cultured pituitary cells was not inhibited. This is evidenced by detection of intracellular LH using immunocytochemistry. Changes in the intensity of LH staining within the cells in dependence of different GnRH concentrations were not observed, but a significant increase LH secretion in pituitary cells was measured. Antarelix had no effect on basal LH secretion. After co-incubation of pituitary cells with Antarelix and GnRH, Antarelix blocked the GnRH-stimulated LH secretion with a maximal effect of 10^{-4} M, but the staining of immunoreactive intracellular LH was detected at approximately the same level compared to the pituitary cells treated with exogenous GnRH alone, reflecting its usefulness [86].

Antarelix

15.30 OTHER GnRH ANTAGONISTS

There are many other GnRh Antagonists which are commercially available. Few important antagonists for clinical purposes include Detirelix, Ramorelix, Prazarelix and Teverelix (Fig. 15.2). They are primarily used in assisted reproduction to control ovulation. The drug works by blocking the action of GnRH upon the pituitary, thus rapidly suppressing the production and action of LH and FSH. They are used in fertility treatment to prevent premature ovulation that will result in harvesting of eggs that are too immature for application in in vitro fertilization.

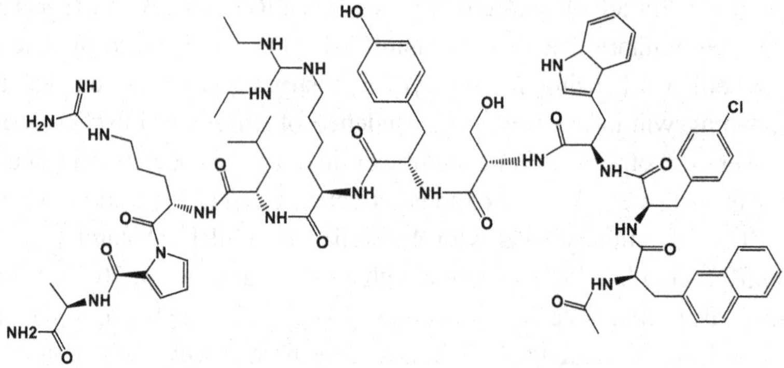

Detirelix

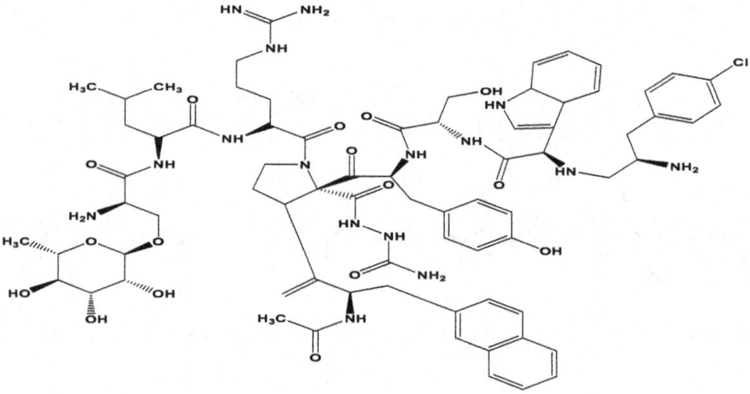

Ramorelix

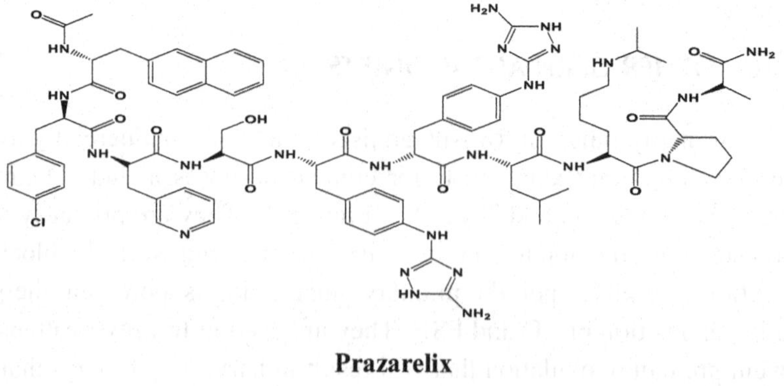

Prazarelix

FIGURE 15.2   Continued

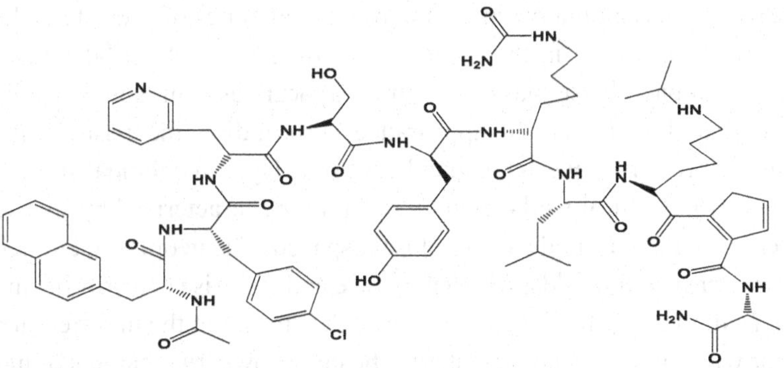

Teverelix

FIGURE 15.2   Structure of some GnRH antagonists.

15.31   GnRH ANTAGONISTS AND PROSTATE CANCER

Prostate cancer is the second leading cause of death across the globe in men after lung cancer. This disease is more prevalent among Negro males, in whom it tends to be more aggressive and progressive, leading to advanced disease. The incidence of cancer increases with age near around 50 years and reaches a peak incidence around 75 years [87]. Autopsy reports indicate that many men above 80 years have latent foci of prostate cancer. The traditional treatment for prostate cancer includes radiotherapy, hormonal therapy, cryosurgery, chemotherapy, radical prostatectomy and receptor antagonism [88, 89]. Prostate is a small secretory tubulo alveolar organ comprising of 20 to 30 separated glands, which opens into the urethra, is primarily responsible for maintenance of sperm motility, clotting of semen and lysis of coagulum. The etiologic influences responsible for carcinoma of the prostate are still not known completely. As with nodular hyperplasia, its incidence increases with age and endocrine changes of old age are related to its origin. Support for this lies in the inhibition of these tumors that can be achieved with orchiectomy. Neoplastic epithelial cells, like their normal counterparts, possess androgen receptors, which would suggest that they are responsive to these hormones. This shows that androgens are required for the maintenance of the prostatic epithelium which in turn also reflects the role of hormone in malignancy. Genetic influences are also involved since blacks are more affected than whites [89, 90].

Prostate carcinoma begins in the peripheral zones of prostate and usually arises anywhere in the gland, in multiple foci, which later fused to form a compact dense mass. The tumor appears as firm, somewhat gritty textured and yellowish in appearance surrounding the tissue. Fibrous stroma is present between the glands. A single layer of cuboidal epithelium is present in acini of well differentiated tumor characterized by smaller in appearance than normal and are closely spaced. The neoplastic epithelium may be thrown into folds, which may fuse and give rise to a cribriform pattern. Malignant epithelial cells infiltrate the stroma in the undifferentiated tumor without any gland formation. The cell shows true features of malignancy [91, 92]. Prostate cancer spreads through lymphatics and veins by direct extension. Local spread occurs in the region of seminal vesicles and base of urinary bladder. Metastases to the regional lymph nodes results in the vascular spread. Hematogenous spread occurs by osseous metastases is a classic example [93]. X-Ray examination detects metastatic lesions in the bone which are diagnostic features of prostate cancer. In axial skeleton, it may be osteoblastic or osteoclastic (destructive).

Four clinical stages are defined (see Table 15.3). Stage A tumors are asymptomatic and discovered on histologic examination of prostatectomy specimens. Stage A_1 lesions are lethal in only a small percentage of patients. Stage A_2 lesions are ominous, leading to death with distinct metastases. Stage B prostate cancers are palpable by rectal examination, but due to peripheral location and small size, they do not encroach on urethra. Patients with stage C and D have symptoms like dysuria, slow urinary stream and urinary retention. Some patients in stage D may present initially with bone pain produced by osseous metastases [87, 89].

Since, prostate tumor proliferation generally obstructs the flow of urine, particularly, the lower urinary tract, an ultrasound scan and serum creatinine level determination the accessibility of the urinary tract [92, 93]. A simple X-Ray of the pelvis and lumbar spine is beneficial to identify osteosclerotic metastases, which reflects the prime evidence of prostatic malignancy. Radioisotope bone scan (RBS) is also used to judge distant metastases in patient. Two specific biomarkers are present, namely, prostatic acid phosphatase (PAP) and prostate specific antigen (PSA) are of value in the diagnosis and management of prostate cancer [88]. Both are produced by normal as well as neoplastic prostatic epithelium. Elevated

TABLE 15.3 Description of the Stages in Prostate Cancer

Stage	
A	Incidental or Clinically unsuspected cancer, detected in tissue removed for apparently benign disorder
	A_1 Well differentiated lesions occupying less than 5% of respected specimen
	A_2 More than 5% of cancer in respected specimen or poorly differentiated lesion
B	Tumors palpable by rectal digital examination but confined to prostate
	B_1 Tumor confined to one lobe
	B_2 Tumor extending to both lobes
C	Tumors that have extended locally beyond the prostate but not produced clinically evident distant metastases
	C_1 Tumor not involving seminal vesicles
	C_2 Extensive periprostatic spread with involvement of seminal vesicles
D	Tumors with distant metastases
	Patients who are presumed to have stage A, B, or C disease clinically
	D_1 but are found to have pelvic lymph node metastases at surgery or by cytologic examination of aspirate
	D_2 Clinical evidence of osseous or distant visceral spread

serum levels of prostatic acid phosphatase (PAP) are present in patients whose tumor has extended beyond the capsule or metastasized. Elevated blood levels of prostate specific antigen (PSA) occur in association with localized or advanced prostate cancer. However, serum levels are also raised in benign prostate cancer, but to a lesser extent compared to previous. But, both of the tumor markers are having its own importance in judgment of disease progression. Immunochemical localization of these markers is also helpful for deciding whether a metastatic tumor originated in prostate. PSA analysis is useful for monitoring response to treatment and disease progression. >100 ng/mL is an indicator of distant bone metastases [89, 90, 92].

Prostatic Cancer is sensitive to steroid hormone and can be treated by hormone depletion only. The management of disease involves surgery, androgen suppression drug therapy or chemotherapy drugs and radiotherapy. The category of drugs which may be used for treatment includes; androgen suppressing drugs, gonadotrophin releasing hormone blockers, alkylating agents such as nitrogen mustard, cyclophosphamide or 5-fluorouracil. Radiotherapy by ^{89}Strontium is used for palliation, but applied complementary for pain control [94, 95]. Prostate carcinoma is not too dangerous if detected early and continued therapeutics and follow-up is concerned. Late identification or subject left untreated for longer periods may proves fatal in course. Treatment at any stage is now possible, but life expectancy or 10 year survival rate is 60–75% as reported. If metastases are present then only 10% survival is reported.

The spread and proliferation of prostate cancer largely depends on the availability of testosterone. The testosterone is converted to dihydrotestosterone by 5α- reductase which is responsible for growth of cancer. Before the Anti-androgens come in regular medical practices castration was the sole method to prevent growth of cancer. In castration, removal of testes leads to further increase in pituitary content and secretions of FSH and LH and hypothalamic lesions prevent this rise. Anti-Androgens were used for the treatment of cancer, by providing testosterone scarcity, but with advancement of research potentiality of Western countries, newer molecules tend to develop which leads to halt the disease progression [88, 92]. Gonadotrophin Releasing Hormone Antagonist (GnRHA) is the newer class of drug for disease treatment that binds with gonadotrophin releasing hormone (GnRH) receptors in the pituitary and blocks the interaction gonadotrophin releasing hormone, thus blocking LH and FSH leads to a periodic blockade of testosterone release from testes, causing prevention of disease proliferation. This can be illustrated in Fig. 15.1, which depicts the steroid feedback pathway. In this pathway, testosterone is inhibits LH secretion by acting directly on the anterior pituitary and by inhibiting the secretion of GnRH from the hypothalamus. Inhibin which is secreted by Sertoli cells, acts directly on the anterior pituitary to inhibit FSH secretion. An inhibition at central level (GnRH) will lead to blockade of peripheral components (FSH, LH) which further inhibits the testosterone release; the main culprit for prostate cancer. Therefore this approach is newer and safer trend and is now prescribed by clinician globally for treatment of carcinoma [89, 90, 92].

15.32   REPLACEMENT OF ANTIANDROGENS BY GNRH ANALOGS

Treatment with the use of hormone-blocking drugs for advanced pros-tate cancer is called antiandrogens. The goal of antiandrogen therapy is to block the effect of testosterone and dihydrotestosterone (DHT) on andro-gen receptors (AR). They are temporarily being used to delay the tumor enlargement, but not a true cure for cancerous growth. Antiandrogens are used primarily at early stages of prostate cancer. Due to the selectivity of nonsteroidal moieties such as flutamide, bicalutamide and nilutamide (Fig. 15.3) to get attached exclusively to AR, they are however referred to as pure antiandrogens [96, 97]. These compounds do not have affinity for gonadotropin, estrogen and progesterone receptor. Thus, considered to be superior than steroidal antiandrogens such as cyproterone, oxendolone and megesterol (given in form of acetate salt), in terms of selectivity, pharma-cokinetic profile and specificity to receptor. Non-steroidal antiandrogens are metabolized primarily in liver and are used in multiple drug combina-tions for treating metastatic prostate cancer. Antiandrogens block the bind-ing of DHT at the AR and blocks or diminish the effectiveness of androgens in androgen sensitive tissues. Thus, are effective for hyperplesia or/and neoplasia of prostate gland. Cyproterone, in form of acetate, suppresses

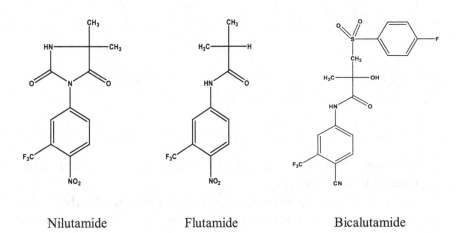

Nilutamide   Flutamide   Bicalutamide

FIGURE 15.3   Continued.

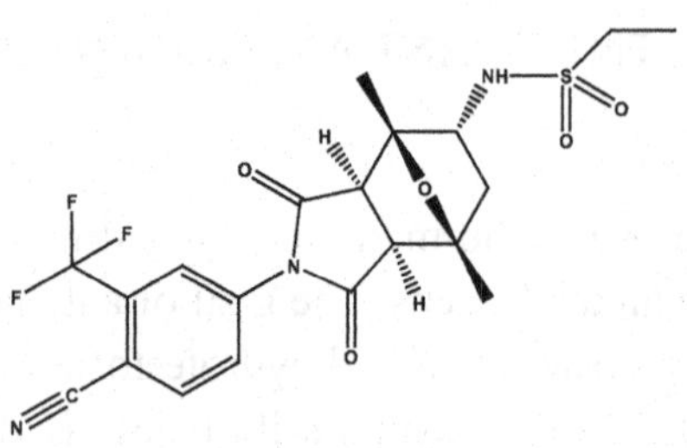

BMS 641988

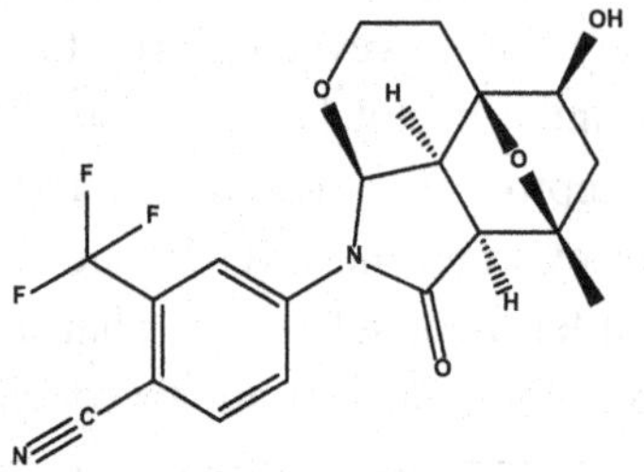

BMS-779333

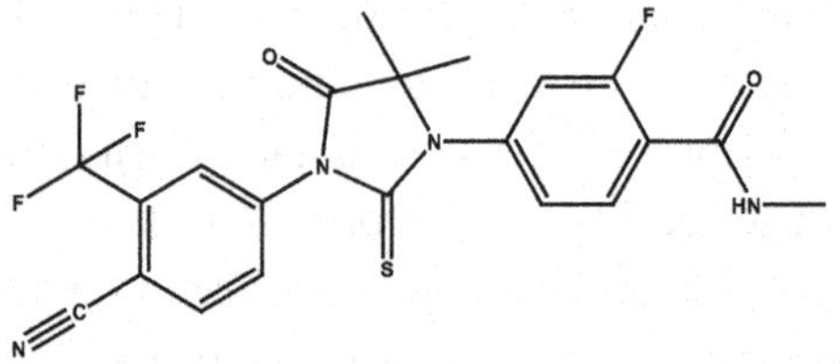

MDV-3100

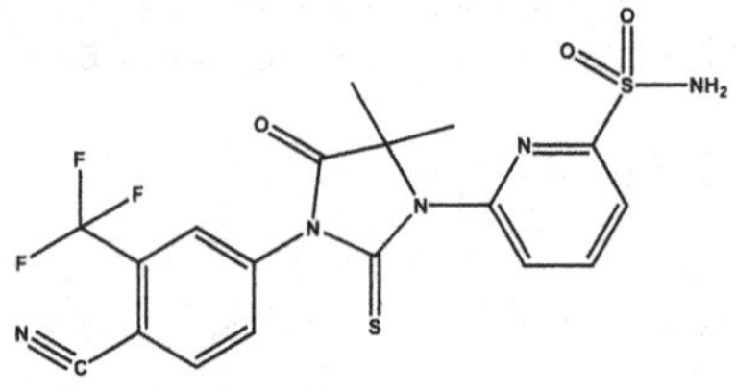

CH5137291

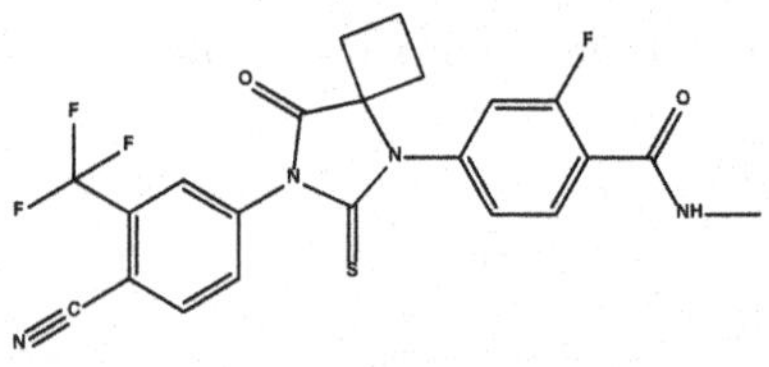

RD162

FIGURE 15.3 Structure of Anti-Androgens.

gonadotropin release and binds with AR. Oxendolone acts by competing for receptor binding sites [98, 99].

Flutamide is administered orally, being completely absorbed from gastrointestinal tract and through CYP1A2 it undergoes extensive metabolism in liver leading to 2-hydroxyflutamide and 3-trifluoromethyl-4-nitroaniline derivatives as primary metabolites. 2-hydroxyflutamide alone forms about 80% of total metabolites produced by liver. Having highest affinity than the parent product, it binds to the receptor to show antagonist effect at low concentration. At high concentration of metabolite, it shows androgenic agonist action also. Flutamide is excreted in urine in form of 3-trifluoromethyl-4-nitroaniline [94, 96]. Bicalutamide is a drug of choice for locally advanced nonmetastatic prostate cancer [100]. Bicalutamide is a racemeate and its (R) enantiomer having 4 times affinity for AR whereas (S) form has no antiandrogenic activity. It is slowly absorbed and has low volume of distribution. After the absorption of drug from the gastrointestinal tract, the (R) form shows stereoselective oxidative metabolism by CYP3A4 in the liver [96, 97]. Nilutamide is a nitroaromatic hydantoin derivative of flutamide used for treatment of metastatic prostate carcinoma in men by its competitive antagonist action on AR. It is absorbed completely after oral administration. One of the methyl group attached to hydantoin ring is stereoselectively hydroxylated to chiral metabolite, which is further metabolized to carboxylic acid metabolite by oxidation. The nitro group was reduced to the amine and hydroxylamine moieties by nitric oxide synthesis [101, 102]. The list of Pharmacokinetic profile of Antiandrogens is described in Table 15.4.

TABLE 15.4 List of Antiandrogenic Drugs with Pharmacokinetic Profile

	Bicalutamide		Flutamide		Nilutamide	
logD (pH-7)	4.9		3.7		3.3	
Oral Bioavailability %	80–90		>90		-	
Onset of Action (Weeks)	8–12		2–4		1–2	
Duration of Action	8 days		3 months – 2.5 years		1–3 months	
Protein Binding %	96		94–96		80–84	
Time to Peak Conc. (hrs.)	31		2–3		1–4	
Elimination Half-life (hrs.)	6		8		40–60	
Cytochrome Isoform	CYP3A4		CYP1A2		CYP2C	
Active Metabolites	none		2-hydroxy		yes	
Excretion %	Urine	Feces	Urine	Feces	Urine	Feces
	34	43	~28	<10	62	<10
cLogP	4.9±0.7		3.7±0.4		3.3±0.6	

15.33 CONCLUSION

The fast development during the last few years underscores our initial statement that GnRHa is a viable alternative to HCG for triggering of final oocyte maturation. GnRHa trigger is safer, patient friendly and offers several physiological advantages over HCG trigger. Although the most optimal luteal-phase support after GnRHa trigger is still being explored, the time has come to question the automatic HCG trigger concept and to move forward with thoughtful consideration of the needs and comfort of patients, specifically in terms of OHSS prevention. GnRH Antagonist offer clinicians a rational first-line hormonal monotherapy option for the management of advanced prostate cancer. These new therapeutic agents not only provide a survival benefit but also show potential for reversing hormonal resistance in metastatic CRPC, and thus redefining hormonally sensitive disease. The development of hormone-resistant prostate cancer (castration-resistant prostate cancer, CRPC) remains the key blockade in successful long-term management of prostate cancer. The release of testosterone, the key hormone is suppressed rather than blockade of active form from binding to the androgenic receptors.

KEYWORDS

- aminoacids
- Benign Prostatic Hyperplasia (BPH)
- cancer
- clinical Importance
- hormone (GnRH)
- testosterone levels

REFERENCES

1. Hayden, C. GnRH analogs: applications in assisted reproductive techniques. Eur J Endocrinol. (2008), 159, S17–S25.

2. Hillensjo, T., LeMaire, W. J. Gonadotropin releasing hormone agonists stimulate meiotic maturation of follicle-enclosed rat oocytes in vitro. Nature. (1980), 287, 145–146.

3. Humaidan, P., Papanikolaou, E. G., Kyrou, D., Alsbjerg, B., Polyzos, N. P., Devroey, P., Fatemi, H. M. The luteal phase after GnRH-agonist triggering of ovulation: present and future perspectives. Reprod. Biomed. Online. (2012), 24, 134–141.

4. Minaretzis, D., Jakubowski, M., Mortola, J. F. Gonadotropin- releasing hormone receptor gene expression in human ovary and granulosa-lutein cells. J. Clin. Endocrinol. Metab. (1995), 80, 430–434.

5. Singh, P., Krishna, A. Effects of GnRH agonist treatment on steroidogenesis and folliculogenesis in the ovary of cyclic mice. J. Ovarian Res. (2010), 3, 26.

6. Maheshwari, A., Gibreel, A., Siristatidis, C. S., Bhattacharya, S. Gonadotrophin-releasing hormone agonist protocols for pituitary suppression in assisted reproduction. Cochrane Database. 2011.

7. Porter, R. N., Smith, W., Craft, I. L., Abdulwahid, N. A., Jacobs, H. S. Induction of ovulation for in vitro fertilization using buserelin and gonadotropins. Lancet. (1984), 2, 1284–1285.

8. Reh, A., Krey, L., Noyes, N. Are gonadotropin-releasing hormone agonists losing popularity? Current at a large fertility center. Fertil. Steril. (2010), 93, 101–108.

9. Smitz, J., Devroey, P., Camus, M., Deschacht, J., Khan, I., Staessen, C., Van Waesberghe, L., Wisanto, A., Van Steirteghem, A. C. The luteal phase and early pregnancy after combined GnRH-agonist/HMG treatment for superovulation in IVF or GIFT. Hum. Reprod. (1988), 3, 585–590.

10. Smitz, J., Ron-El R., Tarlatzis, B. C. The use of gonadotrophin releasing hormone agonists for in vitro fertilization and other assisted procreation techniques: experience from three centers. Hum. Reprod. (1992), 7, 49–66.

11. Humaidan, P., Kol, S., Papanikolaou, E.G. On behalf of the 'The Copenhagen GnRH Agonist Triggering Workshop Group.' GnRH agonist for triggering of final oocyte maturation: time for a change of practice? Hum. Reprod. Update. (2011), 17, 510–524.

12. Humaidan, P., Papanikolaou, for example, Kyrou, D., Alsbjerg, B., Polyzos, N.P., Devroey, P., Fatemi, H.M. The luteal phase after GnRH-agonist triggering of ovulation: present and future perspectives. Reprod. Biomed. Online. (2012), 24, 134–141.

13. Fauser, B. C., Devroey, P. Reproductive biology and IVF: ovarian stimulation and luteal phase consequences. Trends Endocrinol. Metab. (2003), 4, 236–242.

14. Shapiro, B. S., Daneshmand, S. T., Garner, F. C., Aguirre M., Thomas, S. Gonadotropin-releasing hormone agonist combined with a reduced dose of human chorionic gonadotropin for final oocyte maturation in fresh autologous cycles of in vitro fertilization. Fertil. Steril. (2008), 90, 231–233.

15. Humaidan, P., Bungum, L., Bungum, M., Yding Andersen, C., 2006. Rescue of corpus luteum function with peri-ovulatory HCG supplementation in IVF/ICSI GnRH antagonist cycles in which ovulation was triggered with a GnRH agonist: a pilot study. Reprod. Biomed. Online 13, 173–178.

16. Humaidan, P. Luteal phase rescue in high-risk OHSS patients by GnRHa triggering in combination with low-dose HCG: a pilot study. Reprod. Biomed. Online. (2009), 18, 630–634.

17. Humaidan, P., Ejdrup Bredkjaer, H., Westergaard, L. G., Yding Andersen, C. 1500 IU human chorionic gonadotropin administered at oocyte retrieval rescues the luteal

phase when gonadotropin-releasing hormone agonist is used for ovulation induction: a prospective, randomized, controlled study. Fertil. Steril. (2010), 93, 847–854.

18. Radesic, B., Tremellen, K. Oocyte maturation employing a GnRH agonist in combination with low-dose hCG luteal rescue minimizes the severity of ovarian hyperstimulation syndrome while maintaining excellent pregnancy rates. Hum. Reprod. (2011), 26, 3437–3442.

19. Kol, S., Humaidan, P., Itskovitz-Eldor, J. GnRH agonist ovulation trigger and hCG-based, progesterone-free luteal support: a proof of concept study. Hum. Reprod. (2011), 26, 2874–2877.

20. Hoff, J. D., Quigley, M. E., Yen, S. S. Hormonal dynamics at midcycle: a reevaluation. J. Clin. Endocrinol. Metab. (1983), 57, 792–796.

21. Itskovitz, J., Boldes, R., Levron, J., Erlik, Y., Kahana, L., Brandes, J. M. Induction of preovulatory luteinizing hormone surge and prevention of ovarian hyperstimulation syndrome by gonadotropin- releasing hormone agonist. Fertil. Steril. (1991), 56, 213–220.

22. Gonen, Y., Balakier, H., Powell, W., Casper, R. F. Use of gonadotropin-releasing hormone agonist to trigger follicular maturation for in vitro fertilization. J. Clin. Endocrinol. Metab. (1990), 71, 918–922.

23. Wilson, A. C., Meethal, S. V., Bowen, R. L., Atwood, C. S. Leuprolide acetate: a drug of diverse clinical applications. Expert Opin. Investig. Drugs. (2007), 16(11);1–13.

24. Mitchell, M. A. Leuprolide Acetate. Semin Avain Exot Pet. (2005), 14(2):153–155.

25. Abouelfadel, Z., Crawford, E. D. Leuprorelin depot injection: patient considerations in the management of prostatic cancer. Ther Clin Risk Manag. (2008), 4(2):513–526.

26. Chardonnens, D., Sylvan, K., Walker, D., Bischof, P., Sakkas, D., Campana, D. Triptorelin acetate administration in early pregnancy: case reports and the review of literature. Eur J Obstet Gynecol Reprod Biol. (1998), 80, 143–149.

27. Labrie, F., Dupont, A., Belanger, A., Lachance, R. Flutamide eliminates the risk of disease flare in prostatic cancer patients treated with Luteinizing Hormone-Releasing Hormone agonist. J. Urol. (1987), 138, 804–806.

28. Boucekkine, C., Blumberg-Tick, J., Roger, M., Thomas, F., Chaussain, J. L. Treatment of central precocious puberty with sustained-release triptorelin. Arch Pediatr. (1994), 1(12):1127–1137.

29. Rahhal, S., Eugster, E. Histrelin implant for the treatment of precocious puberty. Pediatric Health. (2008), 2(2):141–145.

30. Lewis, A. K., Eugster, E. Experience with the once-yearly Histrelin (GnRHa) subcutaneous implant in the treatment of precocious puberty. Drug Des Devel Ther. (2009), 3;1–5.

31. Rahhal, S., Clarke, W. L., Kletter, G. B., Lee, P. A., Neely, E. K., Reiter, E. O., Saenger, P., Shulman, D., Silverman, L., Eugster, E. A. Results of a second year of therapy with the 12 month histrelin implant for the treatment of central precocious puberty. Int J Pediatr Endocrinol. (2009), 2009, 812517.

32. Wagner, R. A., Piche, C. A., Jochle, W., Oliver, J. W. Clinical and endocrine responses to treatment with desorelin acetate implants in ferret. Am J Vet Res. (2005), 66, 910–914.

33. Schoemaker, N. J., van Deijk, R., Muijlaert, B., Kik, M. J., Kuijten, A. M., de Jong, F. H., Trigg, T. E., Kruitwagen, C. L., Mol, J. A. Use of a gonadotropin releasing hormone agonist implant as an alternative for surgical castration in male ferrets (Mustela putorius furo). Theriogenology. (2008), 70, 161–167.

34. Grosset, C., Peters, S., Peron, F., Fiquera, J., Navarro, C. Contraceptive effect and potential side effects of deslorelin acetate implant in rats (Rattus norvegicus): preliminary observations. Cancer J Vet Res. (2012), 76(3):209–214.

35. Hackshaw, A., Baum, M., Fornander, T., Nordenskjold, B., Nicolucci, A., Monson, K., Forsyth, S., Reczko, K., Johansson, U., Helena Fohlin, Valentini, M., Sainsbury, R. Long Term Effectiveness of Adjuvant Goserelin in Premenopausal Women With Early Brest Cancer. J Natl Cancer Inst. (2009), 101, 341–349.

36. Dixon, A. R., Robertson, J. F., Jackson, L., Nicholson, R. I., Walker, K. J., Blamey, R. W. Goserelin (Zoladex) in premenopausal advanced breast cancer: duration of response and survival. Br J Cancer. (1990), 62, 868–870.

37. Kotake, T., Usami, M., Akaza, H., Koiso, K., Homtna, Y., Kawabe, K., Aso, Y., Orikasa, S., Shimazaki, J., Isaka, S., Yoshida, O., Hirao, Y., Okajima, E., Naito, S., Kumazawa, J., Kanetake, H., Saito, Y., Ohi, Y., Ohashi, Y. Goserelin acetate with or without Antiandrogen or Estrogen in the treatment of Patients with Advanced Prostate Cancer: A Multicenter, Randomized, Controlled Trial in Japan. Jpn J Cancer Res. (1999), 29(11):562–570.

38. Gonadotropin releasing hormone agonists: Expanding vistas. Ind J Endocrinol Metab. (2011), 15(4):625–628.

39. Porter, R. N., Smith, W., Craft, I. L., Abdulwahid, N. A., Jacobs, H. S. Induction of Ovulation for in vitro fertilization using Buserelin and Gonadotropins. The Lancet. (1984), 324(8414):1284–1285.

40. Shapiro, B. S., Daneshmand, S. T., Garner, F. C., Aguirre, M., Hudson, C. Comparison of 'triggers' using leuprolide acetate alone or in combination with low-dose human chorionic gonadotropin. Fertil. Steril. (2011), 95, 2715–2717.

41. Babayof R., Margalioth, E. J., Huleihel, M., Amash, A., Zylber- Haran, E., Gal M., Brooks, B., Mimoni, T., Eldar-Geva, T. Serum inhibin, A., VEGF and TNF-alpha levels after triggering oocyte maturation with GnRH agonist compared with HCG in women with polycystic ovaries undergoing IVF treatment: a prospective randomized trial. Hum. Reprod. (2006), 21, 1260–1265.

42. Engmann, L, DiLuigi, A., Schmidt, D., Nulsen, J, Maier, D., Benadiva, C. The use of gonadotropin-releasing hormone (GnRH) agonist to induce oocyte maturation after co-treatment with GnRH antagonist in high-risk patients undergoing in vitro fertilization prevents the risk of ovarian hyperstimulation syndrome: a prospective randomized controlled study. Fertil. Steril. (2008), 89, 84–91.

43. Imbar, T., Kol, S., Lossos, F., Bdolah, Y, Hurwitz, A., Haimov-Kochman, R. Reproductive outcome of fresh or frozen- thawed embryo transfer is similar in high-risk patients for ovarian hyperstimulation syndrome using GnRH agonist for final oocyte maturation and intensive luteal support. Hum. Reprod. (2012), 27, 753–759.

44. Orvieto, R. Intensive luteal-phase support with oestradiol and progesterone after GnRH-agonist triggering: does it help? Reprod. Biomed. Online. (2012), 24, 680–681.

45. Orvieto, R., Rabinson, J., Meltzer, S., Zohav, E., Anteby, E., Homburg, R. Substituting hCG with GnRH agonist to trigger final follicular maturation – a retrospective comparison of three different ovarian stimulation protocols. Reprod. Biomed. Online. (2006), 13, 198–201.

46. Benadiva, C., Engmann, L. Intensive luteal phase support after GnRH agonist trigger: it does help. Reprod. Biomed. Online. (2012), 25, 329–330.

47. Griffin, D., Benadiva, C., Kummer, N., Budinetz, T., Nulsen, J., Engmann, L. Dual trigger of oocyte maturation with gonadotropin-releasing hormone agonist and low-dose human chorionic gonadotropin to optimize live birth rates in high responders. Fertil. Steril. (2012), 97, 1316–1320.

48. de Mouzon, J., Goossens, V, Bhattacharya, S, Castilla, J. A., Ferraretti, A. P., Korsak, V., Kupka M., Nygren, K. G., Andersen, A. N. Assisted reproductive technology in Europe, 2007, results generated from European registers by ESHRE. Hum. Reprod. (2012), 27, 954–966.

49. Braat, D. D., Schutte, J. M., Bernardus, R. E., Mooij, T. M., van Leeuwen, F. E. Maternal death related to IVF in the Netherlands 1984–2008. Hum. Reprod. 25, 1782–1786. Cerrillo, M., Rodrý'guez, S., Mayoral, M., Pacheco, A., Martý'nez- Salazar, J., Garcia-Velasco, J.A., 2009. Differential regulation of VEGF after final oocyte maturation with GnRH agonist versus hCG: a rationale for OHSS reduction. Fertil. Steril. (2010), 91(4):1526–1528.

50. Bonduelle, M. L., Dodd, R., Liebaers, I, Van Steirteghem, A., Williamson, R., Akhurst, R. Chorionic gonadotrophin-beta mRNA, a trophoblast marker, is expressed in human 8-cell embryos derived from tripronucleate zygotes. Hum. Reprod. (1988), 3, 909–914.

51. Humaidan, P., Papanikolaou, E. G., Kyrou, D., Alsbjerg, B., Polyzos, N. P., Devroey, P., Fatemi, H. M. The luteal phase after GnRH-agonist triggering of ovulation: present and future perspectives. Reprod. Biomed. Online. 2012a;24, 134–141.

52. Humaidan, P., Kol, S., Engmann, L., Benadiva, C., Papanikolaou, E. G., Andersen, C. Y. Should Cochrane reviews be performed during the development of new concepts? Hum. Reprod. 2012b;27, 6–8.

53. Kol, S. Luteolysis induced by a gonadotropin-releasing hormone agonist is the key to prevention of ovarian hyperstimulation syndrome. Fertil. Steril. (2004), 81, 1–5.

54. Devroey, P., Polyzos, N. P., Blockeel, C. An OHSS-free clinic by segmentation of IVF treatment. Hum. Reprod. (2011), 26, 2593–2597.

55. Griesinger, G., Schultz, L., Bauer, T., Broessner, A., Frambach, T., Kissler, S. Ovarian hyperstimulation syndrome prevention by gonadotropin-releasing hormone agonist triggering of final oocyte maturation in a gonadotropin-releasing hormone antagonist protocol in combination with a 'freeze-all' strategy: a prospective multicentric study. Fertil. Steril. (2011), 95, 2029– 2033.

56. Youssef, M. A., Van der Veen, F., Al-Inany, H. G., Griesinger, G., Mochtar, M. H., Aboulfoutouh, I., Khattab, S. M., van Wely, M. Gonadotropin-releasing hormone agonist versus HCG for oocyte triggering in antagonist assisted reproductive technology cycles. Cochrane Database Syst. Rev., CD008046, 2011.

57. Humaidan, P., Polyzos, N. P. (Meta)analyze this: systematic reviews might lose credibility. Nat. Med. (2012), 18, 1321.

58. Papanikolaou, E. G., Pozzobon, C., Kolibianakis, E. M., Camus, M., Tournaye, H., Fatemi, H. M., Van Steirteghem, A., Devroey, P. Incidence and prediction of ovarian hyperstimulation syndrome in women undergoing gonadotropin-releasing hormone antagonist in vitro fertilization cycles. Fertil. Steril. (2006), 85, 112–120.

59. Hernandez, E. R., Gomez-Palomares, J. L., Ricciarelli, E. No room for cancelation, coasting, or ovarian hyperstimulation syndrome in oocyte donation cycles. Fertil. Steril. (2009), 91 (4):1358–1361.

60. Kol, S., Humaidan, P. LH and FSH surges for final oocyte maturation: sometimes it takes two to tango. Reprod. Biomed. Online. (2010), 21, 590–592
61. Van Loenen, A. C. D., Huime, J. A. F., Schats, R., Hompes, P. G. A., Lambalk, C. B. GnRH Agonist, Antagonist & Assisted Conception, Semin Reprod Med. (2002), 20(4):349–364.
62. Cavagna, M., de Almeida Ferreiera Braga, D. P., Borges, E. The effect of GnRH analogs for pituitary suppression on ovarian response in repeated ovarian stimulation cycles. Arch Med Sci. (2011), 7(3):470–475.
63. Lemke, T. L., Williams, D. A., Roche, V. F., Zito, S. W. Foye's Principles of Medicinal Chemistry. Baltimore: Lippincott Williams & Wilkins; 2012.
64. Koechling, W., Hjortkjaer, R., Tankó, L. B. Degarelix, a novel GnRH antagonist, causes minimal histamine release compared with cetrorelix, abarelix and ganirelix in an ex vivo model of human skin samples. Br J Clin Pharmacol. (2010), 70(4), 580–587.
65. Shore, N. D. Experience with degarelix in the treatment of prostate cancer. Ther Adv Urol. 2013, 5(1);11–24.
66. Rick, F. G., Block, N. L., Schally, A. V. An update on the use of degarelix in the treatment of advanced hormone-dependent prostate cancer. Onco Targets Ther. (2013), 6, 391–402.
67. Sonesson, A., Rasmussen, B. B. In vitro and in vivo human metabolism of degarelix, a gonadotropin-releasing hormone receptor blocker. Drug Metab Dispos. (2013), 41(7), 1339–1346.
68. Ruzhynsky, V., Whelan, P. Management of a patient with locally advanced prostate cancer with degarelix: a case report. Can J Urol. (2013), 20(3):6808–6810.
69. Australian Public Assessment Report for Degarelix, Department of Health & Aging, Australian Government, May 2010.
70. Liu, K. K. C., Sakya, S. M., O'Donell, C. J., Flick, A. C., Li, J. Synthetic approaches to the 2009 new drugs. Bioorganic & Medicinal Chemistry. (2011), 11, 1136–1154.
71. Debruyne, F., Bhat, G., Garnick, M. B. Abarelix for injectable suspension: first-in-class gonadotropin-releasing hormone antagonist for prostate cancer. Future Oncol. (2006), 2(6):677–96.
72. Kirby, R. S., Fitzpatrick, J. M., Clarke, N. Abarelix and other gonadotrophin-releasing hormone antagonists in prostate cancer. BJU Int. (2009), 104(11):1580–1584.
73. Tombal, B. New treatment paradigm for prostate cancer: Abarelix initiation therapy for immediate testosterone suppression followed by a luteinizing hormone-releasing hormone agonist. BJU Int. (2012), 109(6):E16
74. Klingmüller, D., Schepke, M., Enzweiler, C., Bidlingmaier, F. Hormonal responses to the new potent GnRH antagonist Cetrorelix. Acta Endocrinol (Copenh). (1993), 128(1):15–8.
75. Lai, Q., Hu, J., Zeng, D., Hu, J., Cai, F., Yang, F., Chen, C., He, X., Yu, Q., Zhang, S., Xu, J. F., Wang, C. Y. Assessing the optimal dose for Cetrorelix in Chinese women undergoing ovarian stimulation during the course of IVF-ET treatment. Am J Transl Res. (2013), 6(1):78–84.
76. Siejka, A., Schally, A. V., Barabutis, N. The Effect of LHRH Antagonist Cetrorelix in Crossover Conditioned Media from Epithelial (BPH-1) and Stromal (WPMY-1) Prostate Cells. Horm Metab Res. (2013), 9, 81–89.

77. Kåss, A., Førre, O., Fagerland, M., Gulseth, H., Torjesen, P., Hollan, I. Short-term treatment with a gonadotropin-releasing hormone antagonist, cetrorelix, in rheumatoid arthritis (AGRA): a randomized, double-blind, placebo-controlled study. Scand J Rheumatol. 2013.
78. Trachtenberg, J., Emerging Pharmacologic Therapies for Prostate Cancer, Rev Urol. (2001), 3(3): S23–S28.
79. European Public Assessment Report, European Medicines Agency, January 2009.
80. Festuccia, C., Dondi, D., Piccolella, M., Locatelli, A., Gravina, G. L., Tombolini, V., Motta, M. Ozarelix, a fourth generation GnRH antagonist, induces apoptosis in hormone refractory androgen receptor negative prostate cancer cells modulating expression and activity of death receptors. Prostate. (2010), 70(12):1340–1349.
81. http://www.news-medical.net/news/20120815/Spectrum-Pharmaceuticals-initiates-enrollment-in-Phase-2b-ozarelix-study-for-prostate-cancer.aspx
82. Frattarelli, J. L., Hillensjö, T., Broekmans, F. J., Witjes, H., Elbers, J., Gordon, K., Mannaerts, B. Clinical impact of LH rises prior to and during ganirelix treatment started on day 5 or on day 6 of ovarian stimulation. Reprod Biol Endocrinol. (2013), 11, 90.
83. Sukcharoen, N., GnRH Antagonists: an Update, Thai Journal of Obstetrics and Gynaecology. (2000), 12, 71–76.
84. http://www.merck.com/product/usa/pi_circulars/g/ganirelix/ganirelix_pi.pdf
85. Deghenghi, R., Boutignon, F., Wüthrich, P., Lenaerts, V. Antarelix (EP 24332) a novel water soluble LHRH antagonist. Biomed Pharmacother. (1993), 47(2–3):107–110.
86. Bellmann, A., Schneider, F., Kanitz, W., Nürnberg, G., Tiemann, U. Effect of GnRH and its antagonist (Antarelix) on LH release from cultured bovine anterior pituitary cells. Acta Vet Hung. (2002), 50(1):79–92.
87. Kumar, V., Cotran, R. S., Robbins, S. L. Basic Pathology. Philadelphia: W B Saunders Company; 1992.
88. Ganong, W. F. Review of Medical Physiology. Connecticut: Appleton & Lange Publishers; 1995.
89. Edwards, C. R. W., Bouchier, I. A. D., Haslett, C., Chilvers, E. Davidson's- Principle & Practice of Medicine, London: Educational Low-Priced Books Scheme; 1995.
90. Sembulingam, K., Sembulingam, P. Essentials of Medical Physiology. New Delhi: Jaypee Brothers Medical Publishers Private Limited; 2006.
91. Singh, I. Textbook of Human Histology. New Delhi: Jaypee Brothers Medical Publishers Private Limited; 2006.
92. Colledge, N. R., Walker, B. R., Ralston, S. H. Davidson's-Principle and Practice of Medicine. Philadelphia: Elsevier Publications; 2010.
93. Sinard, J. H. Outlines of Pathology. London: Elsevier Publications; 2005.
94. Brunton, L., Parker, K., Blumenthal, D., Buxton, I. Goodman & Gilman's Manual of Pharmacology and Therapeutics. San Francisco: The McGraw-Hill Companies Inclusive Limited; 2008.
95. Ritter, J. M., Lewis, L. D., Mant, T. G. K., Ferro, A. A Textbook of Clinical Pharmacology and Therapeutics. London: Hodder Arnold Limited; 1999.
96. Rang, H. P., Dale, M. M., Ritter, J. M., Flower, R. J. Rang and Dale's Pharmacology. Philadelphia: Elsevier Publications; 2007.

97. Tripathi, K. D. Essentials of Medical Pharmacology. New Delhi: Jaypee Brothers Medical Publishers Private Limited; 2008.

98. Brawer, M. K., Crawford, E. D., Labrie, F., Mendoza-Valdes, A., Miller, P. D., Petrylak, D. P. Androgen Deprivation and Other Treatments for Advanced Prostate Cancer. Rev Urol. (2001), 3(2): S59–S68.

99. Wilding, G., Chen, M., Gelmann, E. P. Aberrant response in vitro of hormone-responsive prostate cancer cells to antiandrogens. Prostate (1989), 14, 103–11.

100. Kelly, W. K., Slovin, S., Scher, H. I. Steroid hormone withdrawal syndrome. Urol Clin North Am. (1997), 24, 421–428.

101. Kelly, W. K., Scher, H. I. Prostate specific antigen decline after antiandrogen withdrawal: The flutamide withdrawal syndrome. J Urol. (1993), 149, 607–608.

102. Veldscholte, J., Ris-Stalpers, C., Kuiper GGJM, Jenster, G., Berrevoets, C., Claassen, E., van Rooij, H. C. J., Trapman, J., Brinkmann, A. O., Mulder, E. A Mutation in the ligand binding domain of the androgen receptor of human LNCaP cells affects steroid binding characteristics and response to antiandrogens. Biochem Biophys Res Commun. (1990), 173, 534–538.

INDEX

Printed in the United States
by Baker & Taylor Publisher Services